Chronopathologies

Chronopathologies

Time and Politics in Deleuze, Derrida, Analytic Philosophy, and Phenomenology

Jack Reynolds

LEXINGTON BOOKS
Lanham • Boulder • New York • Toronto • Plymouth, UK

Published by Lexington Books
A wholly owned subsidiary of The Rowman & Littlefield Publishing Group, Inc.
4501 Forbes Boulevard, Suite 200, Lanham, Maryland 20706
http://www.lexingtonbooks.com

Estover Road, Plymouth PL6 7PY, United Kingdom

British Library Cataloguing in Publication Information Available

Library of Congress Cataloging-in-Publication Data
Reynolds, Jack, 1976-
  Chronopathologies : time and politics in Deleuze, Derrida, analytic philosophy, and phenomenology / Jack Reynolds.
     p. cm.
  Includes bibliographical references and index.
  ISBN 978-0-7391-3281-4 (cloth : alk. paper)
  1. Time. 2. Political science--Philosophy. 3. Deleuze, Gilles, 1925-1995. 4. Derrida, Jacques. 5. Analysis (Philosophy) 6. Phenomenology. I. Title.
  BD638.R46 2012
  115--dc23

                                                                    2011038018

Printed in the United States of America

Contents

Abbreviations

Full details of these texts can be found in the bibliography, and unless otherwise noted, references are to standard English translations of these texts.

DR	Deleuze, *Difference and Repetition*
LS	Deleuze, *The Logic of Sense*
A	Derrida, *Aporias*
OT	Derrida, *On Touching*
PF	Derrida, *Politics of Friendship*
R	Derrida, *Rogues*
BT	Heidegger, *Being and Time*
PP	Merleau-Ponty, *Phenomenology of Perception*
TJ	Rawls, *A Theory of Justice*

Acknowledgments

I would like to acknowledge the financial support of the Australian Research Council (2005-2008), which in awarding me a Discovery Grant on this topic allowed me the time to write this book while teaching at La Trobe University and the University of Tasmania respectively. Thanks are also due to those institutions, and more particularly to all of my colleagues at each, particularly James Chase and Jon Roffe. Some of the material in this book was prepared while James and I were co-writing *Analytic Versus Continental: Arguments on the Methods and Value of Philosophy* (Acumen 2010), and the first three chapters of this book hence owe a lot to various conversations with James and the process of writing that other manuscript. The odd expression may even be his, although I have tried to ensure that this is not the case. Equally significant in relation to the later chapters on phenomenology and poststructuralism has been Jon Roffe, my colleague at both institutions and postgraduate student at one. Although he, like James, would not endorse all of the arguments given here, our dialogic disputations have been vital to this manuscript and I thank him for helping me to be a more careful reader of Deleuze than I might otherwise have been, and for showing me that Deleuze is a first-rate philosopher notwithstanding some of the objections about his work that I put forward in this book. Other philosophers who have offered important advice and feedback on various parts of this book include: James Williams, Paul Patton, Philipa Rothfield, Miriam Bankovsky, Sean Bowden, Ashley Woodward, Ricky Sebold, Sherah Bloor, Nick Trakakis, Leonard Lawlor, Christopher Macann, Janna Thompson, Simon Glendinning, Søren Overgaard, Jana Wilson and others at Lexington Books, Rowman and Littlefield. I'd also like to recognize the many anonymous reviewers of my work over the past few years, and especially the one who suggested some helpful structural changes in relation to this book. As always I am indebted to my partner, Jo, and to my children, Rosa and Penelope. Finally, I would like to thank and acknowledge the editors of the following journals for allowing me to reproduce previously published material here, albeit often in substantially reconfigured forms: *Philosophical Forum, Deleuze Studies, Symposium, Theory and Event, Philosophy and Social Criticism, Sophia, Parrhesia, International Journal of Philosophical Studies, Acumen* (the details of these publications can be found in the bibliography).

1

Introduction: The Politics of Time

In the late 1980s, US economist Jeremy Rifkin claimed that, "a battle is brewing over the politics of time."[1] He felt that the pivotal issue of the twenty-first century would be the question of time and who controlled it. In this book I will suggest that a battle over the politics (and philosophy) of time is also what is at stake in the differences between three major currents of contemporary philosophy: analytic philosophy, poststructuralist philosophy (the work of Gilles Deleuze and Jacques Derrida will be our focus) and phenomenological philosophy. No doubt these "umbrella" terms that group philosophers can be, and frequently are, invoked superficially. After all, each of these nominated trajectories is itself a plural beast, having changed significantly over time, and there are also borderline philosophers who trouble any such neat typology. This includes philosophers like Emmanuel Levinas, who sits somewhere between phenomenology and poststructuralism, as well as "postanalytic" philosophers like John McDowell and Robert Brandom who draw on many of the usual suspects associated with continental philosophy, like G. W. F. Hegel, Immanuel Kant, and even Martin Heidegger. Nonetheless, despite such complications I think that there are some real philosophical differences at stake between these three theoretical trajectories, and hope to be able to proffer some useful generalizations about each in this book. In particular, I will suggest that either avowed or tacit philosophies of time serve to define representatives of each of these groups and also guard against their potential interlocutors, not least because of an associated normativity that is grounded in an order of priority and a metaphilosophical assessment of the value and purpose of philosophy, and the methods appropriate to those aims. There is hence not likely to be many full-scale battles between these different philosophical approaches, particularly between the heavily policed boundaries between so-called analytic and continental philosophy; more probable is the persistence of minor skirmishes and perhaps something akin to the cold war, if we are to stick to the military analogies favored by economists and politicians alike. But, by bringing the temporal differences between these three philosophical trajectories to the fore, and showing both their methodological presuppositions and their ethico-political consequences and implications (rather than have them recede as unexplained and uninterrogated axioms), this book will begin a long overdue dialogue on each of these three traditions' respective strengths and weaknesses. In fact, I hope to show that there are systemic temporal problems that afflict each, especially the poststructuralist and analytic traditions, and point to the necessity of a middle-way that does not treat the "living-present" as an epiphenomenon to be explained away as either a transcendental illusion (and as a reactive force that is ethically problematic), or as a subjective/psychological experience that is not ultimately real. Phenomenology is not immune from these temporal biases, either, as I indicate in the discussion of Heidegger in chapter 12 and in various remarks throughout the book on Derrida's deconstruction of Edmund Husserl. But I will argue that such problems are not essential to phenomenology in a manner that jeopardizes my overall project of arguing for a reevaluation of the phenomenological living-present, and more generally the pragmatic temporality associated with bodily coping, as a counterbalance to some of the temporal biases at the heart of analytic philosophy and poststructuralism respectively.

But before my arguments in that regard can be presented, a few preliminary reflections are required in order to indicate some of the links that obtain between time, normativity, and the ethico-political. This is because in even thinking that there is a meaningful conjunction between time, history and politics that warrants philosophical exploration, I am already laying my cards on the table, disclosing that I am not an analytic philosopher, for whom any such conjunction is unlikely to be considered part of the *raison d'être* for philosophy. To put the point less slavishly and less in terms of an identity bestowed from outside, it seems to me that a positive concern with the conjunction of such themes is one of the core family resemblance criteria for being a "continental" philosopher. This generalization need not entail that for the odd case this may seem like a false claim since, when applied to issues regarding the unity of a given philosophical tradition, Ludwig Wittgenstein's non-essentialist idea of overlapping family resemblances denies the need to come up with an exhaustive list of necessary and sufficient conditions that apply to each of the usual suspects associated with that tradition.[2] That said, there are few obvious counter-examples in the twentieth century to this admittedly still vague thesis regarding a continental "temporal turn"– perhaps structuralism – but I remain agnostic about any stronger claim that it is, in fact, a necessary and sufficient condition.

Nor need this generalization about the continental preoccupation with the intersections of time, history and politics mean that the various different forms of continental philosophy are thus effaced, since how to understand this conjunction is variously construed by individual philosophers and by particular groups of philosophers: hermeneuticists, psychoanalysts, existentialists, phenomenologists, poststructuralists, speculative realists, etc. As Elizabeth Grosz suggests:

Each of the three temporal modalities (past, present, and future in all of their conjugative complexities) entail presumptions regarding the others that are often ill- or unconsidered: how we understand the past, and our links to it through reminiscence, melancholy or nostalgia, prefigures and contains corresponding concepts about the present and the future; the substantiality or privilege we pragmatically grant to the present has implications for the retrievability of the past and the predictability of the future; and, depending on whether we grant to the future the supervening power to rewrite the present and past, so too we must problematize the notions of identity, origin, and development.[3]

And certainly there is no agreement in continental philosophy as to the appropriate answer to these and other issues, nor to the precise nature of the relationship between what David Hoy calls the times of our lives and the time of the universe (the "objective" time of physicists).[4] Indeed, part 3 of this book will be substantially oriented around a major fracture in contemporary continental philosophy between the conceptions of time and transcendental philosophy of the phenomenological and the poststructuralist traditions. On the other hand, despite the existence of such philosophical fractures, there is arguably also a methodological agreement of sorts among the vast majority of continental philosophers, which is that starting with the supposition of the ultimate truth of objective time is the wrong way to go (perhaps excepting Quentin Meillassoux).[5] It is typically held that such a methodological starting point means that the time of our lives will not be able to be adequately reconstructed, but it is maintained that if we adopt the reverse procedure, and start from the temporality of our lives, we can adequately explain objective time. Heidegger certainly argues that we cannot understand objective time without first examining the existential time of *Dasein*, as Hoy notes.[6] And such a move is not dissimilar to that made by Henri Bergson and various more recent thinkers, whatever the concrete differences in their actual accounts of temporality. Moreover, in different ways all of the phenomenologists and poststructuralists seek to avoid a conception of time that we might associate with common sense and the natural attitude (i.e., time is a series of instants that is readily measured by clocks), and none have a deferential relation to science in which the truth about time is the physicist's conception and our own experience of time is relegated to the subjective or the psychological, and hence (often) claimed not to be of genuine philosophical import. This enduring interest in the relationship between time, history, and politics is one important marker, among others, which helps to provide a loose philosophical identity to that motley crew that is sometimes sloppily called continental.[7] That said, in my view it also has some kind of diagnostic privilege over other family resemblance features, whether they be methodological or topical, because the endorsement (and rejection) of various different philosophical methods is partly bound up with their success (and failure) in illuminating the relationship between time and politics. Consider the following methods: dialectics, transcendental reasoning post-Kant (where the synthetic a priori is temporalized), genealogy, hermeneutical and psychoanalytic techniques, Heidegger's destructive retrieve (and Derridean deconstruction), the Frankfurt School style critique of modernity, as well as the general wariness of aligning philosophical method with either common sense or a deferential relationship to the findings of the sciences, indexed to the present, and so on (at least one of the last two characterizes a central aspect of the meta-philosophy of most analytic philosophers). From Husserl's genetic phenomenology, to Bergson's *durée* and the use of intuition as a method which is claimed to put us inside rather than outside time, to Heidegger's *Being and Time*, time and method have been central to continental philosophy at least since the start of the twentieth century and, to a lesser extent, since the nineteenth century as we will see. Taken together, these methods also ensure that sustained textual engagement, and a concern with culture and history, undergird large parts of contemporary continental philosophy. To differing extents, all of these continental trajectories insist on the conceptual and historical presuppositions of theoretical frameworks.

This is a manner of proceeding that is distinct from some (perhaps many) of the norms and methods of analytic philosophy, most notably those associated with the linguistic turn and the manner in which analysis is often thought to reveal underlying structures that are ahistorical,[8] as well as the insistence on defining truth as independent of any and all justifications (and hence historical processes of justification), and also, perhaps, the more general analytic concern with argument and rationality in which deductively formalized arguments constitutes a regulative ideal of sorts for philosophical practice, and which might be contrasted with what Simon Glendinning calls the "non-argumento-centric" modes of doing philosophy.[9] Hence the common charge regarding the alleged atemporality and ahistoricality of analytic philosophy. And we know that Quine advocated paying no attention to the history of philosophy and Gilbert Harman reputedly had a note on his door at Harvard University that proclaimed:

"JUST SAY NO TO THE HISTORY OF PHILOSOPHY"

While this is perhaps not an affliction that characterizes the majority of analytic philosophers today, as Brandom protests on the opening page of *Tales of the Mighty Dead*, atemporal tendencies in analytic philosophy certainly remain, even if we find this divide running through the middle of the movement— this is what Peter Strawson calls the Homeric struggle—rather than between it and the continental tradition. However, despite this complication, I will suggest in the chapters that comprise part 1 of this book that some of the less atemporal methods in analytic philosophy (methods that give pre-theoretic opinion, intuitions, or common sense, a central role) typically instantiate a temporal presentism of sorts that is also significantly different from much of what takes place in continental philosophy. Indeed, while we will see that poststructuralist philosophers frequently accuse phenomenologists of being committed to a prioritizing of the living-present, the "now," and invested in a more general metaphysics of presence (Derrida makes both of these charges against Husserl in *Speech and Phenomena*), unlike the presentist tendencies in analytic philosophy, phenomenology remains dubious about the philosophical status of appeals to intuition, pre-theoretic opinions, and common sense. Indeed, this is precisely what any phenomenological reduction must first bracket.

It is worth highlighting a connected issue here, which is the near ubiquity of references to the transcendental and forms of transcendental reasoning within continental philosophy, and the relative scarcity of both in analytic philosophy, excepting some philosophers like John McDowell, Charles Taylor, and Donald Davidson who have been labeled "postanalytic." This term, which became part of the philosophical lexicon in the late 1980s, refers to philosophers who have started out within the analytic mainstream but have been perceived to have "jumped ship" by questioning certain core presuppositions of analytic philosophy. It is sometimes also thought to include philosophers like Wittgenstein and Richard Rorty. Without being able to adequately define postanalytic philosophy here, we might note that the use of transcendental reasoning seems fairly central to the ascription of nonanalyticity by analytic philosophers, and perhaps for the continental philosophers who are increasingly attracted to the work of these postanalytic figures. Without being able to justify this exhaustively here, it seems to be more than simply a coincidence that the work of McDowell and Davidson that depends upon the transcendental is comparatively ignored by the standard-bearing journals of the analytic tradition, whereas other ideas that do not so depend are much more readily integrated and discussed: arguably, this is why the early Davidson is part of the analytic pantheon and the later Davidson is far less significant.[10] There are various reasons for this comparative neglect, not least the fact that influential worries about transcendental reasoning were expressed by Barry Stroud, Stephan Körner, and others in the 1960s in response to Strawson's neo-Kantian use of them in the "descriptive metaphysics" of *Individuals*.[11] Without detailing these issues until the next chapter, we might simply note here that transcendental reasoning can come into conflict with both the analytic respect for common sense and its commitment to what James Chase and I call methodological empiricism – basically, the philosophical desire to either minimize one's footprint outside empirically respectable turf, or to seek empirical respectability.[12]

While very few continental philosophers believe in synthetic a priori reason in the manner described by Kant in his *Critique of Pure Reason*, some form of transcendental reasoning is close to ubiquitous, notwithstanding that what one means by the transcendental is significantly reconfigured by phenomenology, and the genealogical-historicist turn, as well as the linguistic pragmatics of Karl-Otto Apel, Jürgen Habermas, and others. Concerns about the status of transcendental reasoning certainly exist for continental philosophers, but continued creative use persists, and there is no general agreement that transcendental argumentation is especially problematic. In fact, it is more commonly claimed, and it is certainly frequently implied, that a transcendental dimension is of the essence of philosophy. Any philosophical activity that does not reflect on its own conditions of possibility is naïve, or pre-critical, and the sometimes pilloried continental enquiries into the "problem of modernity" are but one way of attempting to reflect on the conditions of contemporary philosophical discourse, subjectivity, and cultural life more generally. Analytic philosophers, on the other hand, might wonder whether transcendental reasoning can survive—as a distinctive argument form—without the synthetic a priori. Does it thus reduce to other argument forms, including inference to the best explanation, or coherentist devices like reflective equilibrium, with the notion of "ground" being entirely divested of its metaphysical implications? There is no consensus in contemporary continental philosophy as to the correct answers to these questions about the fate of the transcendental. Indeed, there is also a significant tension between two contemporary versions of transcendental philosophy—the neo-Kantian version oriented toward justification of principles (in Apel, Habermas, etc.) and the phenomenological version oriented toward clarification of meaning, and in which the "ground" is just what essentially makes any given practice, experience, etc., what it is.[13] Poststructuralism presents a third trajectory, in that we are presented with allegedly necessary complications for either of the above two projects, as well as, especially in Deleuze's work, a new and positive account of transcendental philosophy in which he attempts to provide actual conditions—albeit not standard empiricist ones—for difference and the new. While the terminology of transcendental philosophy may not be there in the work of Michel Foucault in his genealogical period, it is arguable that a similar kind of structure—having a perspective on a perspective from within—is still characteristic of his philosophy, even though it is also true that in a certain sense Foucault inverts Kant by showing that what appears to be necessary is in fact contingent and hence historicizing the a priori.[14]

Throughout these many and varied reinventions of transcendental reasoning, the philosophical task typically remains one of reflecting on one's position as a philosopher, and there is a strong normative desire for any account of experience to provide a nondogmatic account of its own possibility. This typically involves reflecting on one's own time, conditions for that time, and latent forces within it. In his well-known essay "Nietzsche, Genealogy, History," Foucault hence claims that two traditions in modern philosophy come out of Kant's work: "an analytic of truth" and "an ontology of present reality," which he also calls, "a genealogy of ourselves." As Foucault puts it:

> In his great critical work Kant laid the foundations for that tradition of philosophy that poses the question of the conditions in which true knowledge is possible and, on that basis, it may be said, that a whole stretch of modern philosophy from the nineteenth century has been presented, and developed as the analytics of truth. But there is also in modern and contemporary philosophy another type of question, another kind of critical interrogation . . . The other critical tradition poses the question: what is our historical present? What is the present field of possible experiences? . . . It seems to me that the philosophical choice confronting us today is this: one may opt for a critical philosophy that will present itself as an analytic philosophy of truth in general, or one may opt for a critical thought that will take the form of an ontology of ourselves, an ontology of the present.[15]

Foucault's diagnosis is helpful, since almost all of the usual suspects associated with continental philosophy are interested in what Foucault calls an ontology of the present, along with the consequent methodological need to reflect upon the background conditions for any given form of reflection. As Mark Sacks phrases a related point, successful transcendental reasoning involves situated thought, which necessarily involves forms of self-reflection regarding the relationship between given propositional content and the thinking of such contents: as such, psychological and historical matters are not radically disassociated from truth, although one obviously needs to be careful articulating this relationship.[16] Another way of putting Sacks' point about transcendental philosophy and situation might be to consider the notion of the background that is a focus of much of continental philosophy in its various forms: we might think here of talk of horizons, life-world (*Lebenswelt*), motor-intentionality, transcendental fields, absorbed coping, context, and the interest in the social, historical and psychological conditions for certain kinds of philosophical utterance and ways of living. For Husserl, Heidegger, Merleau-Ponty, and others, central to their phenomenological reflections is an attempt to make more perspicuous previously inarticulated background conditions.

When it comes to thematizing the ethico-political and normativity more generally, continental philosophers also invariably invoke time, and this often depends on forms of transcendental philosophy and temporal orders of priority. This is partly due to the vast influence of Heidegger, including his direct association in *Being and Time* of "vulgar" or ordinary time (i.e., time understood as fundamentally what is tracked by clocks) and inauthenticity (see §81). Roughly speaking, for Heidegger an inauthentic mode of being-in-the-world is one without temporal unity and an authentic mode of being-in-the-world has temporal unity, a thesis that is subsequently contested by Levinas, Derrida, and Deleuze (as well as others) despite their indebtedness to some other aspects of Heidegger's philosophy of time. But Heidegger didn't come from nowhere in this respect. Other continental philosophers associated time and normativity in the nineteenth century, and we will see in the various chapters that comprise part 2 that Derrida and Deleuze continue to do so. While Hegel is an obvious case, arguing that the task of philosophy ought to be to grasp one's own time in thought, we might also think of Friedrich Nietzsche's revaluation of values in which time is central. Nietzsche argues, for example, that all *ressentiment* is resentment of the present (the "now") and claimed in *Ecce Homo* that the notion of the eternal return of the same was his greatest idea, along with the associated idea of *amor fati*: become what one is. We might also consider Karl Marx, whose rich and varied analyses of the relation between certain modes of production and time (e.g., time-as-measure) remains influential. Søren Kierkegaard's *Concluding Unscientific Postscript* is primarily concerned with the manner in which the genuinely religious life involves a contradiction between temporal existence and eternity, as well as the manner in which the choice, or leap of faith, occurs at an instant in which time (lived time) and eternity are envisaged to intersect. As such, I think that my claim about a "temporal turn," which traces roots in Nietzsche, Marx, Kierkegaard, Kant, and others, is stronger than Simon Glendinning alleges in a recent journal exchange that he and I had on this issue. Glendinning says:

> It may well be that poststructuralists and this new wave have ploughed the furrow of a temporal turn that distinguishes it from most analytic philosophy today, and I explicitly accept in the book that confining the title to the new wave is a coherent and understandable strategy. However, with respect to the post-Kantian line of thinkers who are undoubted contributors to the texts that make up the primary works of Continental philosophy, we should not regard that strategic arrogation as showing up the unity or even "quasi-unity" of a distinctive Continental tradition.[17]

Contrary to Glendinning, I think that reading such post-Kantian figures in relation to a temporal turn is not merely a strategic appropriation of these eighteenth and nineteenth century texts; we can see clear evidence of these claims in the texts of the vast majority of the usual suspects. While such an interpretation is conducted with the benefit of hindsight and involves a relationship to these texts and thinkers that is irremediably altered by concerns that became more prevalent in the twentieth century, not only is this hermeneutically inevitable but it is also precisely why I restrict my claims to a meaningful continental tradition to the early twentieth century and the work of Husserl, Bergson, and Heidegger, alleging that their work constitutes a temporal tipping point, but not a moment of radical rupture from what precedes it.

This desire to contextualize ethico-political problems historically, and to treat how one experiences time as central to the good life (however that is understood) has various consequences. It means that continental philosophers rarely invoke rule following accounts of ethics and politics. Very few, for example, would concur with Brad Hooker, Elinor Mason, and Dale Miller, who in their introduction to *Morality, Rules and Consequences: A Critical Reader*, argue that the task of ethical theory is to come up with a code of rules or principles, that would, ideally (a) allow a decision procedure for determining right action in a particular case, and that could (b) be stated in such terms that any nonvirtuous person could understand it and apply it correctly.[18] Rather, they are instead typically engaged in forms of ethico-political reflection that are closer to virtue ethics, in that the focus is not typically on the consequences of particular acts, nor even their motivating intentions (in the sense of what one deliberately and consciously aimed to accomplish by a given act), but on character and patterns of action over periods of time. With Nietzsche, for example, one key question is: could we wish for our whole life, with everything that happens in it, to occur again without difference?

Moreover, Nietzsche dismisses utilitarianism as an ethics for children and mediocre Englishmen[19] and it is significant that no well-known continental philosopher has ever, as far as I can discern, subscribed to utilitarianism and it has rarely even been discussed seriously. In the analytic tradition, by contrast, it has been the dominant ethico-political position of the twentieth century (as John Rawls acknowledges). Act-utilitarianism is explicitly committed to impartiality as a value and to the principle of choice for a society being the same as for an individual. It is also often characterized by its critics as requiring an implausible calculation of the greatest interest for each and every act and decision. It is atemporal. Of course, not many act-utilitarians still exist, and those who do are often characterized as extremists whose work sacrifices all of our common intuitions and thus tacitly requires a radical reconfiguration of ordinary moral thought. Contemporary utilitarians usually try and moderate their perspective by bringing it more in line with our everyday moral intuitions, with what people currently think is morally just and other intuitive judgments, and hence it becomes a kind of temporal presentism, accommodating itself to the moral intuitions currently held by many (e.g., analytic philosophers) or most people more generally (e.g., the "folk"). Only a minority of utilitarians think that having individual agents consistently calculating what would be the most just course of action is actually in the best interests of maximizing welfare or happiness, because people would get it wrong and thereby produce more harm than good, and because it would preclude living a unified life of "integrity" pursuing a single career within an enduring communal nexus. As Tim Mulgan suggests, most contemporary utilitarians agree that a theory's decision procedure can diverge from its criterion of rightness, and usually utilitarians will argue that if time is short, then either choose the first good enough option that comes along (satisficing) or choose the best option of those you have time to consider (constrained maximization), or follow rules of thumb.[20] An extension of this final position is to advocate rule utilitarianism, which argues that utilitarian judgments should apply to general social rules (or to institutions) but should not be taken to apply at the individual level. It hence avoids the consistently voiced objections to utilitarianism that it is counterintuitive by mitigating act utilitarianism's extreme commitment to impartiality, which means we have the same obligation to an unknown Rwandan as we do to our parents and children, and apparently also licenses the hanging of innocent people to preserve overall well-being.

Equally as dominant a force in contemporary analytic philosophy is liberalism—indeed, the post 1970's renewed analytic interest in ethics and politics stemmed at least partly from Rawls' liberal manifesto, *A Theory of Justice*. Here Rawls insists on the core principle of "justice as fairness" and a rational proceduralism about how to bring this about that distances itself from questions of the good life. He argues, among other things, that we should seek to maximize the situation of those worst off in any particular society (often called the maximin principle). And, at least in *A Theory of Justice*, the resultant principles are meant to hold atemporally (thanks to the test of the "original position," the veil of ignorance), even if a motivating factor for its emergence is the situation of pluralist societies with competing conceptions of the good. Now, just as no well-known continental philosopher has ever subscribed to utilitarianism, surprisingly few have unambiguously subscribed to liberalism (excepting Raymond Aron and the *nouveaux philosophes* who in the 1970s and 80s broke with both Marxism and poststructuralism in France), instead offering something like communitarian critiques of liberalism and its "unencumbered self."

Likewise, it is also worth noting that the major focus of analytic political philosophy is with distributive justice, being a key concern of utilitarians, liberals, and others. Much has been said about this preoccupation with distributive justice by Michael Walzer, William Connolly, Bonnie Honig, and Iris Marion Young,[21] the last three of whom all draw substantially on continental philosophers to mount their criticisms and might be considered as such themselves. Along with the German theorist of recognition, Axel Honneth, they have all sought to supplant this preoccupation with distributive justice with a return to what they consider to be the more fundamental concerns of justice, such as domination, oppression, and power. These themes were, of course, all foregrounded by Hegel's "struggle for recognition," and Marx also criticized any focus on distributive justice as inevitably neglecting what he felt to be more important issues to do with the inequality in the forces of production (and the resultant alienation), and thus tinkering in piecemeal fashion with an unjust society. From all of these diverse perspectives there is a sense in which the contemporary analytic focus on distributive justice is itself a kind of reification of capitalist relations in that, as Young puts it, it is preoccupied with having (i.e., possessing) more than doing. Political values are treated like a static commodity that one either possesses or doesn't, rather than a social and temporal process.[22] There is perhaps also a resultant tendency to treat distributive justice and social justice as coextensive when they should not be. For Young at least, this bias precludes adequate engagement with issues to do with domination and oppression, and the problem is compounded by the fact that moral and political theories in the analytic tradition focus on deliberate and conscious action— that is, on the agent's present aims and preferences. Unintended and unconscious sources of oppression tend to be ignored, as are the social and institutional structures that produce certain kinds of agents and determine certain kinds of distributive patterns.

We consider this in more detail throughout the next few chapters of part 1 of the book, but analytic political philosophy focuses predominantly on synchronic issues, continental philosophy on diachronic issues. Anglo-American political philosophy is typically either atemporal in its assessments of value (e.g., utilitarian calculations, or the alleged neutrality of liberalism), and/or presentist in moderating these orders in a more coherentist manner, say via the influential method of reflective equilibrium and the reliance upon thought experiments which have become endemic to the analytic tradition since about the 1960s. From the perspective of Derrida and Deleuze's work, as we will see, there is a common sense conservatism built into this privilege accorded to the present, and the atemporal preoccupation with reasoning and the calculable threatens to ignore important background considerations and simplify problems in order to attain solutions (or procedures for solutions) that remain at a distance from, or epiphenomenal to, the affective dimensions of the problem itself. Now, there are, of course, various analytic replies to such charges, perhaps most notably that they constitute a form of "para-politics" as Hilary Putnam puts it—that is, a politicized philosophy that sees itself primarily in social and political terms.[23] Illustrating these differences in method, and their topical consequences, will be a key task of this book. Although temporality will be one of the main themes, it is also a methodological leitmotif, in that I will be interested in what these various philosophical methods disclose about our time-embedded status, something that is a particular focus of chapter 3. My contention will be that there is a certain kind of chronopathology—that is, time sickness or theoretical myopia about time—that afflicts each of these three traditions, but especially analytic (political) philosophy and the poststructuralism of Deleuze and Derrida. In some ways my "cure" for this sickness will consist simply in the juxtaposition of each, allowing us to see (at least to an extent) each tradition through the other's eyes, since what becomes clear from this juxtaposition is that intimately linked to the different philosophies of time at work in these two traditions are their different treatments of that which admits of calculation (i.e., is present and numerable) and that which is incalculable (i.e., transcendental and futural). And there is a tendency to place too great an emphasis in either direction—the analytic tradition prioritizing the former, the poststructuralist the latter—and I argue that this has important and problematic ethico-political consequences for both camps in relation to their respective treatments of politics, violence, decision-making, and rationality.

In addition, this book offers, especially in part 3, a series of phenomenological reminders motivated by, and sometimes explicitly grounded in, the work of Merleau-Ponty, and which also selectively draws on aspects of the philosophies of time of Husserl and Heidegger. This is to show two inter-related things: (1). the lack of necessity of the various transcendental claims made by Deleuze and Derrida that use time to legitimate normative orders of priority; and (2). that the analytic and poststructuralist traditions both unjustifiably neglect the importance of what Hubert Dreyfus (drawing on Merleau-Ponty) calls *l'habitude*, a kind of pre-cognitive comportment and embodied coping that, coimbricated with our proprioceptive sense of own bodies, is also fundamental to the constitution of the "living-present" and has its own ethical significance. This minimal proprioceptive sense that human animals, and many non-human animals, have from the beginning (i.e., the earliest stages of foetal life) is simple an unrefined positional awareness of one's limbs and body, but as Merleau-Ponty shows in *Phenomenology of Perception*, it also serves as the basis for the development of a body-schema, habits, and even intelligent skills and learning, as we seek to establish maximum grip, or optimal *gestalt*, within an environment. For Merleau-Ponty, these aspects of bodily motility and perception are the transcendental conditions that ensure that sensory experience has the form of a meaningful field rather than being a fragmented relation to raw sense data. Drawing on such analyses, I argue against Deleuze and Derrida's transcendental accounts of time that in different ways render this living-present as secondary, as well as their normative move that makes of it reactive coping and thus denigrates the movement toward an equilibrium—albeit that is never finally attained—that is involved in our embodied comportment toward the world.

However, my main aim is not to contend that Merleau-Ponty is right *tout court*, and that philosophy has regressed since his untimely death in 1961. Rather, what is ultimately proposed is something like a model of mutual constraint. Each of these theoretical trajectories alerts us to methodological and normative problems in the others surrounding their respective philosophies of time, and my inference to the best explanation is that the various normative injunctions derived from metaphilosophical orders of priority cannot ultimately be sustained. David Wood suggests that violence (both of a conceptual and more empirical nature) is best understood as fundamentally a disease of time, as a "chronopathology."[24] Agreeing with this idea, but also using it against some of the poststructuralist thinkers who likewise deploy it, we will see that ethico-political problems are in store for us when the living-present is understood as self-contained, when we are nostalgic for the past (or seek to return to some origin), and when the future is understood as entirely circumscribed and delimited by the expectations of the present. But it also rears its head when the living-present, the bodily coping and integrative aspects of our experience of time, are thought of as either exhausting the nuances of bodily life (as the phenomenological position sometimes seems to maintain) or are instead treated as epiphenomenal and a transcendental illusion. I will hence suggest that the philosophies of Deleuze and Derrida succumb to their own form of chronopathology in making something like this latter claim, despite their acute insights about this very phenomenon. While Husserl discusses the inevitability of a transcendental pathology, I argue that Deleuze and Derrida both make this mistake in regard to their transcendental (and ethical) valorisation of the futural synthesis of time, a disjunctive synthesis that both understand as pertaining to time, but "time out of joint." Too intent on escaping presentism, their account of the future is insufficiently "embodied and embedded" to borrow a phrase from contemporary cognitive science, and it a priori excludes the ethico-political importance of embodied recuperative time. My focus here will be on Deleuze's rich texts of the late 1960s, and Derrida's work post-1980 because of its more explicit ethico-political concerns. They have been chosen not only because of the influence of their respective philosophies of time, but also because of the way in which their work enables particularly productive conversations with the analytic tradition, and because the sidelining of embodied coping that undergirds their work represents the most sophisticated forms of a position that is shared more widely by poststructuralist philosophers and which this book will seek to challenge.

Part One

**Analytic Philosophy,
Atemporality, and Presentism:
Some Encounters across the *Chunnel***

2

Analytic and Continental Philosophy: A *Contretemps*?[1]

Judging purely by the names bestowed upon them, the distinction between analytic and continental philosophy is rather confused. It simultaneously invokes both a method—analysis—and a geographical location—continental Europe. Even on its own terms, the geographical determination is clearly inadequate, since most so-called continental philosophy is currently practiced in the United States, and Europeans (especially Germans and Austrians) were fundamental to the emergence and development of early analytic philosophy. Despite these definitional problems, however, I think Bernard Williams and Putnam are wrong to conclude that it is a distinction without a difference, and that we are all just doing philosophy with no need for any preceding adjective like continental or analytic.[2] In what follows, I will begin to set up my argument that very different philosophies of time, and associated methodological techniques, serve to define representatives of each of these groups and is the means by which these traditions of philosophy differentiate themselves from their competitors.

To begin with, then, let me offer a patchy history of philosophy of time in the early twentieth century, the period in which the idea of a divide between two ways of doing philosophy became entrenched. Primarily through the work of John McTaggart, Bertrand Russell, Husserl, Bergson, and then slightly later in the work of the logical positivists (especially Rudolph Carnap and Hans Reichenbach) and Heidegger, time came to the forefront of both philosophical attention and the broader social milieu as these philosophical accounts of time were forced into an engagement with physics, particularly Einstein's theory of relativity. Einstein showed, firstly, that time is relative, in that if one travels at close to the speed of light away from their friend, their time will pass considerably slower compared to their friend who remains on Earth. This might seem to call into question the notion of objective or

17

absolute time, but, given other related developments in physics particularly in quantum mechanics, it actually led to the generally accepted idea (by most analytic philosophers and physicists) that rather than time and space being separable we need to think of a single space-time continuum (or block) with four-dimensions, of which time is a one-dimensional subspace. Such a position entails a radical questioning of any primacy accorded to the "now," perhaps best exemplified by the philosophical position termed presentism, where only the present is thought to be ontologically real, and the past and the future are not—e.g., since we can't point to Socrates anywhere, he is not real.[3] In fact, it is usually accepted that four-dimensionalism in physics means that our experience of the now (the immediate present) and the notion of temporal becoming are but subjectively compelling illusions. This constitutes quite a challenge to both common sense conceptions of time and our everyday experience of time.

I will return to the question of whether such reductionism is necessary toward the end of the chapter, but the continental temporal turn reached its tipping point with the work of Bergson, Husserl, and Heidegger. Arguments against any ontological privilege accorded to the "now" and any conception of time based on a succession of instants were also raised by these philosophers, but generally their reflections on time were not couched in terms of the latest developments in physics or mathematics. Certainly, the deferential relation to the findings of the sciences that is evinced by most analytic philosophers is not apparent. Bergson, widely regarded as the most famous philosopher in the world from the turn of the century to World War 1, initiated a very different revolution in thinking about time around the start of twentieth century, illustrating the importance of a non-measurable lived time (*durée*) that is also said to be the proper medium of thinking (via "intuition" rather than the intelligence), and as a result he proposes radical alterations to the way we think about memory. For him, the past remains real and part of all of our experience, existing as what he calls a virtual temporality (a time that is real, but not present); it is not reducible to the linear order that clock-time, or the succession of instants model gives us. This latter view, which many of us hold uncritically, understands time as the chronological succession of an infinite series of "nows" or instants stretching from the future to the past. Why the future first, and the past second? Well, if a priority is given to this present instant, then we know that there are various things that have not yet happened (a *not-yet-now*), which we will come closer to (note the spatial metaphor), which will then happen, and which will then be in the past (a *no-longer-now*). In contrast to such a view, in *Matter and Memory* Bergson refers to a notion of the "pure past" that coexists with the present, and constitutes it; the present, on this view, is a dimension of the past rather than vice versa.

Around the same time, Husserl's rich analyses of internal time-consciousness demonstrated that, rather than conscious life being comprised of a series of punctal "now" moments, there is a necessary intertwining of anticipative "protention" in relation to the future and "retention" of the past. His concern was to describe our experience of time, as well as to delimit some of the non-empirical conditions of possibility for any experience of time. For most phenomenologists, it is this relation between protention, primal impression, and retention, which makes possible the experience of, say, temporal flow. We need not have an actual experience of protention, but we must deduce such features in order to coherently explain our experience of, say, listening to a melody. In other words, these features are claimed to be the transcendental conditions of temporal experience in its myriad forms, and these time-constituting phenomena are not themselves determinable as occurring in the present, past, or future, in the same way as the events that historically happen.

In 1927, Heidegger's *Being and Time* also gave time a transcendental significance, insisting that for *Dasein* (and more generally) a philosophical priority must be accorded to the "not yet," the possible, and the future. (BT §68) In a pragmatic sense, for Heidegger we cannot use chalk in a lecture theater, or meaningfully negotiate a basketball match, without aiming at something in the future (in some general sense). Moreover, he argues that we aim pragmatically at fulfilling certain tasks because of an originary futural ecstasis of time: we have something we want to be or become (this is closely related to what Sartre, in *Being and Nothingness*, calls a fundamental project). But beyond this Heidegger also claims that temporality is a necessary condition for *Dasein*; it is the ground, without which there is no *Dasein* at all. To summarize Heidegger's key claims, he suggests that all "understanding" depends upon the futural ecstasis of time; that "attunement" (or state-of-mind, or mood, depending upon the translation of *Befindlichkeit*) is structured by the past ecstasis of time (throwness, having-been); and that "fallenness" is fundamentally an attachment to present things that denies these other escstases. To put it bluntly, understanding, attunement, and fallenness can hence only be understood through time. Moreover, the temporal structure that Heidegger calls "care" is the basis from which he seeks to derive his philosophical accounts of space and place, (BT § 70) and this privilege that he initially gave to time over space has been influential on much of the ensuing continental tradition, especially French phenomenology and poststructuralism.

Indeed, Heidegger also argues for the need to engage in a destructive retrieval—*Destruktion* ("destructure") and *Abbau* ("unbuild")—of the meaning of Being in relation to the history of philosophy, which is, he famously says, the history of the forgetting of Being. Heidegger's basic claim is that much of the Western philosophical tradition has made the mistake of understanding time as an entity among other entities. This metaphysics of presence has meant that both time and Being have hence been understood almost exclusively as thing(s), as that which is objectively present and "there," and any metaphysical account of "reality" is then constructed on the basis of this understanding. Heidegger argues that there is a resultant prioritization of theoretical ways of knowing and understanding that he terms the present-at-hand (that which can be analyzed as a distanced and apparently neutral observer) rather than the ready-to-hand (that which we use in an engaged and practical manner). Without wanting to explicate Heidegger further at this point, he suggests that this also entails a prioritization of the calculable, numerable, and quantifiable, and a failure to understand time except spatially. For Heidegger, this preoccupation with calculation also has consequences for politics and ethics, even if only very loosely so, in that a concern with the possession of resources and technological control becomes dominant, rather than the less specifiable concerns with, say, communal flourishing, dwelling, and authenticity.

It is important to consider the dialogues that did and did not occur during this formative period of the divide. It is telling that there was very little substantive engagement between the phenomenological/Bergsonian accounts of time and the anglophone (analytic and idealist) discussion of McTaggart's metaphysical concerns with what he called the A-series and B-series of time. Indeed, McTaggart's 1908 paper, "The Unreality of Time" was very influential upon the analytic tradition that was to follow, with the community even today roughly polarizing around two perspectives—presentism and eternalism—that overlap substantially with what McTaggart designated as the A-series and B-series, even if few accepted his conclusion which is captured by the paper's title.[4] Briefly, McTaggart illuminated two ways in which positions in time can be ordered. His A-series of time represented a psychological experience of succession that roughly corresponds to what the phenomenologist might call the natural attitude in regard to time, with certain events being futural, coming to be present, and then moving into the past and the even further past. All of these temporal designations are relative to a given present, from which certain events are seen to be in the future and others in the past. To put this another way, for the A-series positions in time are orderable according to their possession of properties like being two days future, being present, being one day past, and so on. But McTaggart also suggested that positions in time can also be ordered by two-place relations like two days earlier than, one day earlier than, simultaneous with, and so on. This B-series of time—of before, now and after—involves a succession that maintains permanent relations among events, and suggests that temporally tensed sentences (like "I will finish this talk within the allocated time") are not required. Most analytic philosophers agree that modern physics supports this view. Advocates of the A-series are hence forced to either dispute this claim about modern physics, or argue that if it is true then modern physics leaves out something fundamental. This latter approach, although relatively rare in analytic philosophy, seems to be the position that most continental philosophers are tacitly committed to, and this is explicitly the case in Heidegger's work as we will see.

There was, on the other hand, a series of polemical encounters between Russell and Bergson,[5] and one of the major philosophical points of contention between them concerned time. Russell was considerably worried about Bergson's anti-intellectualist valorization of intuition, having once famously quipped, "intuition is at its best in bats, bees, and Bergson."[6] From Bergson's perspective, however, bound up with the method of intuition, and justifying its philosophical deployment, was his understanding of time. Bergson's view can be summarized as arguing that "the real nature of time resides not in its segmented parts but in its given, experiential character as duration: the irreducible, purely qualitative, cumulative flow of a multiplicity of states which forms an indivisible, heterogenous continuum."[7] All attempts to understand time (or, for that matter, life, or existence) by partitioning it into quantifiable items are bound to fail. They are spatializing abstractions that use the "intelligence" to fixed break things down into their subcomponents (what Michael Beaney might call decompositional analysis),[8] but for Bergson time is underivable or irreducible, in that it cannot be analyzed by reference to anything which is nontemporal. And we should note that the real nature of time was thought to pertain to this experiential (and, simultaneously, transcendental) characteristic, not to the four-dimensional continuum advocated by physicists and analytic philosophers. For Bergson, it is the former that is the condition for the latter, rather than the other way around. *Prima facie* this kind of sentiment sits uneasily with analytic philosophy as a method, and, unsurprisingly, Russell was rather antagonistic toward such a view. He was not prepared to accept the qualitative (and ontological) difference between time and space that Bergson insisted upon. And Russell could empirically justify this by pointing to work in physics and mathematics. On a similar basis, J. J. C. Smart, W. V. Quine, and many other analytic philosophers have argued that there is no objective ontological difference in kind between the past, the present, and the future, just as there is no ontological difference between here and there. It would be conceded that "Yes, we thank goodness that the pain is there rather than here, and past rather than present, but these differences are subjective, being dependent on our point of view."[9] It is merely mental perspectives that divide the block into a past part, a present part, and a future part, and that radically distinguish time and space. Partly because of some similar considerations, Russell even felt entitled to proclaim that time was an "unimportant and superficial characteristic of reality."[10] To Bergson, on the other hand, time is closer to ultimate reality, a transcendental condition of the spatial, of change, and of life.

Competing philosophies of time were also part of the background to the polemic between Heidegger and Carnap in the early 1930s that is well analyzed in Michael Friedman's book, *A Parting of the Ways: Carnap, Cassirer and Heidegger*. After all, we have seen that Heidegger's *Being and Time* argues that Being is not an entity and that the question of Being has a vital relation to time. Carnap, Reichenbach and other logical positivists were substantially concerned with the philosophical implications of the theory of relativity, especially in regard to space and time. Carnap wrote a thesis setting out an axiomatic theory of space and time, and Reichenbach became a well-known philosopher of science who wrote many books on the philosophical implications of Einsteinian physics. Both became major figures in the United States and were instrumental in the rise of analytic philosophy there. On the other hand, Heidegger's *Being and Time* gave almost no consideration at all to Einsteinian physics. His later work, where he talks of "time-space" rather than Einsteinian space-time, indicates his feeling that Einsteinian physics (like Newtonian physics, which is directly addressed in *Being and Time*) continued to treat time as a spatial entity (a thing).

While logical positivism as a program has been largely abandoned in analytic philosophy, some of the crucial differences that were apparent at this stage (and earlier in the Bergson-Russell debate) have recurred throughout the twentieth century. Most notably, the analytic tradition harbors reservations about transcendental philosophizing about time, and contains a strong culture of deference to the best findings of the relevant sciences, while the vast majority of continental philosophers, by contrast, typically use transcendental arguments and are either critically/creatively engaged with science (non-deferentially), or are entirely uninterested in it. What this means for the philosophy of time is well-captured by Keith Ansell-Pearson's summary of the Russell-Bergson debate: "the physicist gives us space-time in which time has no independent meaning; but the philosopher holds that space-time is really spatialized time and not time at all. The physicist then retorts that the time of the philosopher is merely phenomenological or psychological."[11] Ansell-Pearson's "physicist" here is a proxy for many analytic philosophers, including Russell, Carnap, Reichenbach, and myriad recent four-dimensionalists. If this characterization is valid, and I think it is, then it seems plausible to maintain that in regard to the philosophy of time something like what Lyotard calls a *differend* continues to bedevil analytic and continental philosophy—that is, a case of conflict between two parties that does not allow for the possibility of a rule or criterion (or even a linguistic vocabulary) by which the dispute may be fairly decided.[12] We have seen the continental interest in the historical presuppositions of concepts and frameworks. Moreover, since Heidegger, continental philosophy has been wary of forms of temporal presentism (including references to immediacy, intuitions, and any postulation of an undivided "now") in a manner that is radically distinct from the analytic metaphysical privileging of it (presentism) or indifference concerning it (four-dimensionalism). Moreover, for most (if not all) continental philosophers the very possibility of philosophizing, thinking, and questioning, requires that one consider the relationship between time and subjectivity,[13] whereas the majority of analytic philosophers evince by their practices that this is not the direction to go. This divide only gets reinforced when metaphysical or transcendental reflections on time become associated with sociopolitical critique, as happens with Marxist philosophers and, in a different way, in the poststructuralist guise of continental philosophy that we will now consider.

Time and Politics

Many of the most famous continental philosophers have seen an intimate relation between philosophies of time and socio-political issues, including Hegel, Marx, and Heidegger. How might we explain or argue for this connection? From a sociological and psychological perspective, one might point to the manner in which capitalism and urban life has caused many of us to feel that time has sped up and accelerated. Many of us feel that we have a kind of time-sickness—that is, the belief that time is running away from us, making us impatient, and that we need more of it, as if it were something that could be added to all of the other things we need more of: money, goods, love, esteem, and the like. In his book, *In Praise of Slow*, Carl Honoré suggests that "the clock is the operating system of modern capitalism, the thing that makes everything else possible—meetings, deadlines, contracts, manufacturing processes, schedules, transport, working shifts."[14] He details all kinds of societies who have responded to this monitoring of time by staging various kinds of slow movements that embrace a different understanding of time—Austria's Society for the Deceleration of Time, Japan's Sloth Club, the Long Now Foundation in the United States. Many have abandoned clocks and radically changed the structure of their working week. These various organizations also point to quite convincing empirical studies that show us that time-pressure can lead to tunnel-vision, which, of course, has political and social consequences in the form of bad decision-making. In fact, in this book I want to claim something similar: that theoretical myopia about time, or to put it another way a reductive or nonpluralistic approach to time, can create philosophical problems and lead to bad methodological decisions which ultimately result in partial philosophical conclusions.

But challenges to this dominant conception of time as involving a series of moments, a linear trajectory that clock-time regulates and subjects to our

control, might also be made on philosophical grounds. In fact, it is this understanding of time as about measurement, and thus as amenable to calculation,
prediction, arithmetical division, and ultimately time-keeping, that all of the
different poststructuralist thinkers challenge, sometimes on ethico-political
grounds but also on philosophical ones. Most frequently, the philosophical
reasons rely on various forms of transcendental argument, which are given to
explain why linear-clock time (or theoretical ideas of time that are dependent
on a clock-like series of moments) are an abstraction from lived time, or are
dependent upon the existence of a past that cannot be recalled and a future
that cannot be anticipated, and incoherent without them. Methodologically
speaking, the prolific use of transcendental arguments within continental philosophy constitutes a major point of difference with analytic philosophy as
we have seen, but to view this as the fundamental metaphilosophical cause of
the analytic-continental divide would be to ignore the fact that many of the
other major continental methods develop and affirm our essential historicity,
as does the conception of reason that is invoked.

While Derrida raises some important questions about the Heideggerian
project and conception of time (and asks is there any concept of time that is
not "vulgar" or metaphysical),[15] Heidegger's criticisms of the metaphysics
of presence has nonetheless been very influential on him. For Derrida, at
least in early texts like *Of Grammatology* and *Speech and Phenomena*, all
metaphysics emphasizes that which *is*, or that which appears, and has forgotten to pay attention to the condition for that appearance. In *Speech and
Phenomena*, Derrida makes clear that this emphasis upon presence of the
Western tradition takes two main forms: something *is* insofar as it is capable
of presenting itself as an objectivity to thought, or the subject *is* insofar as it is
self-present, transparent to itself in the immediacy of a conscious act.[16] Either
way, truth, or meaning, is construed to be a presence without difference from
itself, and from this tacit emphasis upon presence comes the subject-object
dichotomy and various other conceptual oppositions that persist to the present day. Deconstruction performs two strategies in relation to this emphasis
upon presence: firstly, it emphasizes that which makes presence possible,
much as Heidegger also did; secondly, it also looks at what makes presence
impossible, or at least problematic within the structures and claims of a given
text, and Derrida affirms this to a greater extent than Heidegger does. Indeed,
this is why Derrida claims that his neologism *différance*—which conjoins
spatial differing and temporal deferring—is "older than" Heidegger's ontological difference.[17]

Derrida and the other poststructuralists also inherited the importance that
the Heidegger of *Being and Time* accorded to the futural aspects of time (the
"not yet", projecting toward a potentiality) in their significance for *Dasein*'s
being-in-the-world, as well as for any hermeneutic understanding. In their

work, the futural synthesis of time that *interrupts* (rather than *integrates*) is accorded even more importance. This aspect of time, which is argued to be the time of the event, cannot be anticipated, predicted or calculated; it resists, it withdraws. A politics of the future (synonymous with a philosophy of the event), along with a related denunciation of the kind of calculative thinking that understands the event exclusively as "soon to be present" can be located in the work of virtually all of the key poststructuralist thinkers, including Levinas, Lyotard, Foucault, and Negri.[18] While Levinas' philosophy of time is more commonly associated with an emphasis upon an irrecuperable past that obliges us to respond to it, even though we cannot represent it or know it, his position on futurity is an important predecessor to both Deleuze and Derrida's. In *Time and the Other*, for example, Levinas holds that "anticipation of the future and projection of the future, sanctioned as essential to time from Bergson to Sartre, are but the present of the future and not the authentic future; the future which is not grasped, which befalls us and lays hold of us. The Other is the future."[19] In *Totality and Infinity*, he likewise insists on the importance of that which interrupts a Husserlian conception of the "now," with its integration of protentive and retentive aspects along with a primal impression. Developing such insights, Derrida and Deleuze emphasize a conception of the future whose significance lies in the way in which it simultaneously interrupts any "now" or present moment and is a condition of possibility for any event worthy of the name; otherwise there is no possibility of the event but merely a pre-programmed or deterministic outcome. On this view of the future it is fundamentally not amenable to prediction and calculation, unlike analytic philosophy, which arguably tacitly endorses this calculative model in relation to the future in the frequent use made of logical modelling, probability analyses, and the general concern with distributive justice in political philosophy.

But what is wrong with anticipating the future and saying the future will be like this or that? While holding that we cannot avoid anticipating the future, and that we also must try to predict and calculate future consequences for any decision that hopes to be responsible (i.e., to say that we must prepare), Derrida's work intimates that exclusive adherance to these kind of relations to the future can be dangerous. Certainly if such a relationship to the future, a calculative one, becomes dominant in a given milieu, it *can* lead to the absolutist violence of fascism and communism in which in the name of this or that future state of affairs individuals silence or kill those who do not believe in such a future. Less dramatically and more frequently they lead to the minor fascisms of everyday life in which noone dies but conceptual and interpersonal violence nevertheless takes place. This means that Derrida will not name the future. Democracy and justice must instead be understood as "to come" and unable to ever be actualized in a present situation. This trajectory of Derrida's thought is examined in greater detail in chapter 5, but it is worth emphasizing that this futural dimension of time that he refers to has a curious structure in that it intervenes in the time of the present, breaks it open, without ever fully coming to presence, and a similar predicament afflicts our relationship to the past. One cannot, after all, attain to a self-contained present that is immured of the past, and yet this past cannot be presented before us and recognized in its totality. The influence of the past upon our present is largely unspecifiable, despite what some psychoanalysts suggest, and yet they are right in seeing such influence as being pervasive. Borrowing from Levinas, Derrida maintains the thought that there is something akin to an "ethical past," an immersion in the weight of history that is immemorial. This ethical past cannot be represented or recalled, one cannot assume responsibility for it as a totality, and the trick is hence not to see the past as an origin to which one might return. In contradistinction to any temporal presentism, then, Derrida reminds us of both the future and its radical indeterminability, and also the immemorial past that defies conscious memory and representation but nonetheless subsists in the form of traces.

In his analyses of the times of Chronos and Aion in *The Logic of Sense*, Deleuze makes some closely related points. As we will see in detail in chapter 6, he describes two different aspects of time that he also calls the *time of the present*, where the past and future are only dimensions of the present contracted into it, and the *time of the past and future*, where the present is only a dimension of the past and of the future. Chronos is described as the time of actual events where all past and future events are synthesized into an indivisible present akin to Bergson's *durée*. On this view, which is similar to that of Augustine and contemporary analytic metaphysical presentists, no reality can be ascribed to that which *is* not actual, but, unlike such presentists, the present is still extended and has past and future dimensions.[20] For Deleuze, however, the time of Aion is equally real and is described as the disruptive moment in which any given present is always dividing into both the past and the future, such that there is, in fact, no present. Only the past and future truly subsist in time on this view, and Deleuze hence calls Aion the time of the eternal and the time of the pure event. As such, it is the future and the past that affect us; an openness to the future that necessarily resists our calculative entreaties and that immemorial past which cannot be represented as a totality. Another way of putting this point might be to say that if rationality purports to be atemporal, albeit with its claims being expressed in the present-tense (inference and verification are of the present tense), then there is something that will always elide such forms of reason.[21] We might say, as Levinas does, that rationality (and he would also say scientific time) involves a synchronization of the diverse that cannot comprehend diachrony: in this case, the past that subsists and the future that cannot be anticipated but instead disrupts.[22]

This account of a continental "temporal turn" could easily be extended. Sartre's paradoxical conception of consciousness in *Being and Nothingness* (it is what it is not and is not what it is) revolves around his account of time, just as Merleau-Ponty's philosophy of ambiguity in *Phenomenology of Perception* does: "I know myself only in my inherence in time and in the world, that is, I know myself only in my ambiguity" (PP 345). The tradition of critical theory owes a great deal to Walter Benjamin and his Judaic philosophy of messianic time (which also heavily influenced the work of Giorgio Agamben), as well as to Marx and others. In many of its guises (e.g., Herbert Marcuse) critical theory also relies heavily upon psychoanalytic techniques that trace the impact of the past within the present. Hans Gadamer and other hermeneutic philosophers arguably also take this temporal turn, albeit without the poststructuralist valorization of the future as difference, novelty, and the like. For Gadamer, *Verstehen* and phronesis are given greater import, as are our common sense prejudices, which are seen as the ground of any possible understanding. It is also worth noting the existence of some recent important books by Bernard Stiegler, Lyotard, and Éric Alliez on time, and the fact that Alain Badiou's philosophy of the event is also arguably preeminently concerned with a rupture within time.[23] Furthermore, in *The Life of the Mind* Hannah Arendt develops St. Augustine's characterization of humans as "*homo temporalis.*"

While we have obviously run together some widely divergent and competing accounts of time here in this characterization of continental philosophy, it remains the case that there is a preoccupation with time in the work of all of these philosophers and movements. Moreover, time is understood in a manner that pertains more to social life than to Einstein and other developments in physics, along with a certain anti-presentism about time that carries with it an ethico-political inflection. For all of these philosophers, there is an intricate and complicated topology to time, which, if simplified (e.g., treated linearly, spatially) can degenerate into problematic political positions and to violence. To put the point another way, any such misconstrual of time involves a transcendental violence that is often the progenitor of more empirical instances of violence. David Wood, we have seen, specifically understands violence as a disease of time, as a "chronopathology,"[24] and he suggests that what Nietzsche diagnoses as *ressentiment* (a disgust for life that trades in negativity) is fundamentally a taking revenge against the fact that time passes. The major form that this *ressentiment* consists in is by artificially delimiting time and insisting upon the priority of the present, and this could include the invocation of a self-contained intuitive present, or an ahistorical calculative deduction. Ethico-political problems are in store for us when we jealously seek to preserve some feature of the present.[25] Problems are also likely when we are nostalgic for the past or seek to return to some origin, and violence is not far away when the future is circumscribed and delimited by the expectations of the present. Institutions and societies are worthy of critique to the extent that any of these above chronopathologies take on a dominant form. Now, this account of time and its relation to ethico-political matters is unlikely to be taken seriously by many analytic philosophers, possibly on account of the charge that such claims remain psychological points buttressed with transcendental claims that never adequately display their necessity, or that they constitute a kind of parapolitics as Putnam suggests.

Time in Analytic Philosophy and Beyond

While time has been a major area of philosophical inquiry in analytic metaphysics, philosophy of science, and personal identity, no wholesale turn like that which I have associated with continental philosophy ever takes place. An explicit engagement with time is not a mandatory aspect of any genuinely original philosophy, although the less piecemeal and more systematic projects in analytic philosophy must give philosophy of time serious attention. Moreover, such philosophers often envisage themselves as clarifying the physics, which is ultimately the truth about time, or contextualizing and situating it within a framework of other epistemological and metaphysical claims. Most contemporary analytic philosophers are hence four-dimensionalists about time, developing an account of time that supplements post-Einsteinian physics. It would, however, be a simplification to argue that the difference between analytic and continental philosophies of time is reducible without remainder to the issue of the status of the philosopher's role in relation to the findings of physics. There are, after all, plenty of analytic metaphysicians of time who do armchair philosophy about time involving conceptual analysis, and who are not particularly concerned with the latest developments in physics. In fact, there is a split between the four-dimensionalists who bite the physicist's bullet and say that the passage of time (and the experience of the "now") is an illusion, and others (like John Bigelow and Arthur Prior) who would claim it is a Moorean fact and not an experiential illusion, a split that can be traced all the way back to the work of McTaggart. Perhaps it is possible to argue for the compatibility of presentism and relativity/four-dimensionalism as Bigelow has recently done,[26] but such views are yet to be widely accepted. Both of these dominant trajectories will be considered here, but there is also a further issue that we will address. If time and method are bound up with one another according to continental philosophers, and if they are right about postulating this connection, then an obvious question must be asked: is there an implicit philosophy of time underwriting some of the major methodological practices of analytic philosophy? To put it another way, what do the main philosophical methods presuppose about the analytic philosopher's time-embeddedness, for want of a better term? Perhaps some of the methodologies that are used are temporally restricted, reliant on the present even while the content of their avowed philosophical positions is generally stridently anti-presentist or atemporal.

This is, in fact, what I will argue in what follows since, *prima facie*, the main methods associated with continental philosophy are not deployed in the analytic tradition: genealogy, hermeneutics, deconstruction, etc., are not fundamental, and nor is transcendental argumentation. Indeed, analytic philosophy's general scepticism about the value of transcendental arguments (and therefore about forms of first philosophy more generally) is certainly operative in the area of philosophy of time. This methodological wariness partly explains the two main ways in which philosophy of time has been practiced: a deferential relationship to the physicist's understanding of time; and a pursuit of what might be called minimalist metaphysics without transcendental arguments. As Robin Le Poidevin and Murray MacBeath note, it seems that Kant's antinomies (time has a beginning; time has no beginning) are the source of the analytic feeling that a priori arguments for time having a certain structure or topology are doomed to end in failure.[27] Despite this, Heidegger claimed in *Kant and the Problem of Metaphysics* that it is Kant who in fact made a decisive temporal turn, a turn that Heidegger argues is *the* turning point in modern philosophy and one within which his own philosophical project can be situated.[28] In response to this incredulity when it comes to first philosophy and transcendental argumentation, Bill Newton-Smith's book, *The Structure of Time*, argues that the structure of time is a matter that can only be settled empirically; physicists alone can determine, for example, whether or not time is bounded.[29] In a related vein, in *Time's Arrow and Archimedes' Point* Huw Price seeks to reestablish an Archimedean point for knowledge outside of the relativism that seems to be a consequence of the special theory of relativity and our various anthropocentric biases, most particularly the fact that our philosophizing and thinking about time is greatly affected by our own finite status as creatures in time. Price adopts the reverse procedure to Heidegger and attempts to dispel rather than dwell on this paradoxical temporal structure by reinstating an objective atemporality, a view from "nowhen."[30]

When analytic philosophers do metaphysics of time, on the other hand, this is not in the sense that one might think. Rather than venturing into what might be termed "strong" necessary conditions of possibility terrain, the issue of whether time passes, or what the structure of time is, is often understood to be answered via logic and an analysis of what makes tensed statements true. There are also many thought experiments in which one is asked to imagine that someone knows all that there is to know about time, history, and dates, but does not know which date is the present one. Richard Swinburne and Ernest Sosa argue, for example, that this suggests there are temporal subjective facts.[31] Likewise, there are thought experiments like Sidney Shoemaker's "frozen time," and ongoing analyses of time-travel as philosophers attempt to decide if the asymmetry of causation is related to the asymmetry of before and after.[32]

Radical revisions to our common sense understandings of time are coun-
tenanced by many of these philosophers. Nowhere does it seems that a chronological, linear understanding of time as a succession of instants is dominant,
and yet nowhere is there any kind of reflection like that for which continental
philosophers go in for, whether they be phenomenologists or poststructuralists. Can we even situate the continental perspectives that we have considered
in terms of the debate between two views on the nature of temporal reality—presentism and eternalism—that have dominated analytic philosophy?
Both seem to fail to capture what is at stake in continental reflections on time,
whether at the beginning of the twentieth century or the end. Most continental
philosophers since Heidegger want to dispute the philosophical priority of
the present, as the majority analytic eternalist camp does, but also to insist
(unlike eternalists) on temporal becoming, and on ontological distinctions between past, present and future, as well as between time and space. This conjunction of claims has no equivalent in analytic philosophy, perhaps because
it is not readily compatible with physics. Of course, there are some analytic
philosophers who feel that the eternalism and presentism alternatives are unduly restrictive too, including those who maintain (often Wittgensteinians)
that the debate is merely verbal because each side is using the word "real"
in a different sense; the presentist uses it in a tensed sense, whereas the eternalist uses it in an untensed sense.[33] Nonetheless, perhaps due to the practice of conditionalization (maximizing inferential connections so dialogue
and communal progress can be made)[34] much of the debate revolves around
evaluating the pros and cons of two (or at most three) main accounts of time:
presentism, eternalism, and to a lesser extent the "growing universe" theory.
None of these frameworks resemble in either methods or conclusions the
kind of positions and arguments proffered by Deleuze and Derrida, or for that
matter by Husserl and Bergson. Consider, for example, the commonly es-
poused B-series claim that: "when we say that the year 1900 has the property
of being past, all we really mean is that 1900 is earlier than the time at which
we are speaking. On this view, there is no sense in which it is true to say that
time really passes, and any appearance to the contrary is merely a result of the
way we humans happen to perceive the world."[35] This idea that the past might
become part of an omni-temporal space-time block, rather than retain its own
significance, is precisely what the notion of an immemorial past (Levinas and
Derrida), or a virtual past (Bergson and Deleuze), would deny.

There is, likewise, a genuine clash between the phenomenological accounts of time and what analytic philosophers are interested in. In one sense,
this is perfectly understandable. Phenomenology focuses upon describing our
experience of time, which might include the interplay between the three main
temporal dimensions as they are differently experienced in, say, boredom,
anxiety, or listening to a melody (or even as capitalism impacts upon our ex-

perience of each). In addition, phenomenologists delimit some of the nonempirical conditions of possibility for the experience of time. It hence examines subjective experience, and the conditions of possibility for that experience, whereas physicists are concerned with objective or physical time (what is ultimately out there in the world). What are we to make of these very different accounts? Can they both be true? It seems we are faced with a dilemma: pluralism or reductionism?

We can be pluralists, and say that both are right in their own domain (roughly the subjective versus the objective, the phenomenological versus the naturalistic), but this risks a resulting vacuity or lack of philosophical depth, as well as not seeing the ramifications that each view has for the other, in that their must be some interrelation between these different philosophies of time unless we are prepared to countenance a radical dualism—and few analytic or continental philosophers are. Indeed, if physical time and psychological time are two different kinds of time, then extensive commentary is required regarding their relationship to one another, something that is generally not forthcoming in the work of either analytic or continental philosophers. Moreover the contrast between the subjective and objective domains does not quite capture what is at stake, since it ignores the transcendental focus of continental arguments about time. Even phenomenology, for example, is seeking to offer an account of some of the core structural features of subjective experience, rather than merely a subjective account of experience.[36]

Alternatively, we can be reductive and say that one of these two broad philosophies of time derives from the other, the latter of which is more fundamental. Both analytic and continental philosophers tend to choose this option far more often than ecumenical pluralism, although the methods and rationale provided for these orders of priority is very different. How else might we explain the suggestion made by many analytic philosophers that subjective or psychological time is an illusion, and that physical time is more fundamental even though psychological time is discovered first by each of us as we grow out of our childhood, and even though psychological time was discovered first as we human beings evolved from our animal ancestors?[37] How else might we explain the following phenomenological claim from Shaun Gallagher and Dan Zahavi's *The Phenomenological Mind*?

> There is no pure third person perspective, just as there is no view from nowhere. To believe in the existence of such a pure third-person perspective is to succumb to an objectivist illusion. This is, of course, not to say that there is no third-person perspective… but it is a perspective founded upon a first-person perspective, or, to be more precise, it emerges out of the encounter between at least two first-person perspectives; that is, it involves inter-subjectivity.[38]

On such a view, the conditions of subjective time are also the conditions of possibility of inter-subjectivity, which are, in turn, the condition of possibility of the objective time of the physicists. Moreover, for them any account of time that ignores its experiential development in humans, that gives no attention to questions concerning the conditions of possibility of temporal experience, will be one-sided.

Indeed, issues pertaining to the genesis of knowledge and its development are important to all continental philosophers, thus inviting the charge of perpetuating the genetic fallacy in thinking that an analysis of origins has anything to do with the evaluation or justification of present knowledge claims. Sometimes this is perhaps a fair charge, but the reverse is also a risk, what James Williams has termed the anti-genealogical fallacy,[39] an accusation that might be bestowed on any view that thinks that concepts or reason come from nowhere, like manna from heaven, or that we might translate, without significant remainder, our over-determined and messy language with its history of associated concepts into a pure language that overcomes such difficulties. Indeed, contemporary four-dimensionalists are committed to the view that the correct theory of time, the metaphysical truth about time, is ultimately tense-less. As Brad Dowden suggests, for eternalism or the block universe theory:

> The future-tensed sentence, "the Lakers will win the basketball game" might be analyzed as, "The Lakers do win at time t, and time t happens after the time of this utterance." The future tense has been removed, and the new verb phrases "do win" and "happens after" are tenseless logically, although they are grammatically in the present tense.[40]

Yet, *prima facie*, our experiences of time and the way we learn about the world is through and through temporally tensed. This claim might be justified by pointing to the historical presuppositions of concepts that continue to wield influence over our thought (as with genealogical, hermeneutic, and deconstructive techniques), or it might be more phenomenologically grounded in our experience as body-subjects: we navigate in the world through our bodies and their temporal anticipative capacities. Bodily motility and intentionality simply is temporally tensed. For example, the perception of an object is informed by procedural memory (habits) and it also gives us hitherto undisclosed sides, sides that our attention might be directed toward in the future. Of course, a possible rejoinder to this might be that this temporal experience says nothing about whether these subjective phenomena are ultimately real. But, for most continental philosophers, at least from Husserl and Bergson on, such experiences are the transcendental condition of our experience of a world at all, from which scientific analyses are based, and any prior determination of what it is for something to be real inevitably partakes in a metaphysics of presence. Analytic philosophers, on the other hand, might accept that a condition of the development of the sciences is this embodied dimension, but insist that the truth (or metaphysical reality) of what is discovered leaves such genetic questions behind. Does this commit many analytic philosophers to positing a view from nowhere as it was criticized by Thomas Nagel in his book of that name, or more to the point does it commit them to positing an atemporal view from "nowhen," as Huw Price has happily accepted in his book *Time's Arrow and Archimedes' Point*? I cannot adequately resolve this dispute here, but it seems that we have encountered a *contretemps* that helps to explain the rather pervasive nonengagement between representatives of analytic and continental philosophy in the twentieth century. There will, no doubt, be many counter-examples to any thesis like this (perhaps including Willfrid Sellars, the teacher of Brandom, McDowell and others),[41] but this chapter has nonetheless begun to show what is distinctive about the philosophy of time of large parts of continental philosophy and also distinguished this from the philosophy of time that typically takes places in analytic philosophy. Are we confronted here with incommensurable paradigms or stances that do not admit of criteria that might resolve the dispute in a nonquestion begging way? Perhaps, but it is also possible that a more pluralist methodology can help us here, especially one that can find a place for transcendental arguments.

The myriad criticisms raised against such arguments by analytic philosophers cannot be rigorously addressed here, other than to note they generally felt to be dogmatic because of: (1). a presupposed idealist/verificationist premise (Barry Stroud); or (2). the inability to determine the unique applicability of the claimed condition (Stephan Körner). Körner's worry is that they have not (and necessarily could not) rule out other alternative explanations and that they hence amount to a form of dogmatism.[42] Stroud's argument is that the sceptic can always maintain that it is merely necessary that we *believe* in the transcendental condition—say, that there are extramental facts—since this is sufficient to account for the mental experience in question, but it does not show that there *are* any extra-mental facts.[43] There are various possible responses to each of these charges, including that they take transcendental reasoning to depend on the synthetic a priori, or to be aimed at a refutation of the skeptic, neither of which is a fundamental part of contemporary continental usage, which are more typically about conditions of meaning (for phenomenology), or regulative in nature. Moreover, even if Stroud and Körner's objections do highlight a certain fallibility at the heart of transcendental arguments, as it seems to me that they do, such reasoning ought not be too hastily prohibited on the grounds that they are sometimes, or perhaps even frequently, liable to error (i.e., dogmatic claims to necessity, circularity, etc.). Being primarily concerned with error avoidance is a conservative philosophical ambition that does not necessarily coincide with the ambition to seek truth or achieve genuine insight; the risk of error might be eliminated only to be replaced by the risk of banality or obsolescence.[44]

But this argument need not depend on one wholeheartedly agreeing that transcendental arguments can be usefully deployed. After all, it is clear that certain philosophical methods focus upon our affirming our historicity (and our exposure to an unknown future) more than others. While no particular method can immunize us against the risk of philosophical failure, there is good reason to prefer the prospects of a more pluralist methodology that can find a place for genealogical, transcendental and phenomenological analyses, as well as many of the various techniques that have become part and parcel of analytic philosophy. Such a diverse toolbox of methods might ensure that we do not remain enthralled with one temporal modality at the expense of others and that we do not preserve the name truth for any single one of those domains. Such a synthesis is improbable and perhaps it is even impossible at least in the foreseeable future, but we must at least confront the prospect of meta-philosophical dialogue and disagreement. Without it, philosophy of all kinds can become insular and all too confident of expertise in a given narrow domain of academic specialization. Such an attitude is not very philosophical. If learning only occurs, as Deleuze suggests, intermediate between knowledge and non-knowledge (DR 206), let us hope that philosophers of all varieties can, at least from time to time, leave behind their habitual specializations and certitudes.

3

Common Sense and Philosophical Methodology: Some Metaphilosophical Reflections on Analytic Philosophy via Deleuze[1]

In *Discourse on Method*, Descartes famously observed that what is most evenly shared and equally distributed in the world (i.e., common) is good sense. This sense is not precisely defined by him, nor by most of those who have subsequently invoked it, but it seems to encompass our basic reasoning and inferential abilities as well as something closely related to what has come to be called "folk psychology," which refers to our everyday ability to attribute desires, beliefs, and intentionality to other people despite the theoretical possibility of them being but cloaks and springs.[2] Indeed, Descartes argues in his *Meditations* that if we are methodologically careful enough (and with the help of God), this sense will enable us to justify many of our varied claims to knowledge. Judging by some recent accounts, however, this "sense" is neither quite as common (e.g., universal) nor as good (e.g., naturally oriented to truth) as Descartes and many others would have us believe. Whether we consider the continental tradition, starting with the "masters of suspicion" and phenomenology, and continued in different ways by both the structuralists and poststructuralists, or whether we consider the challenge to the analytic epistemological tradition proffered by experimental philosophers like Stephen Stich, this sense has come in for a battering in recent times. Stich and others critique old-fashioned conceptual analysis (and also its residue in the work of Davidson, Dennett, and others) for amounting to an "intuition driven romanticism" because of the assumption that both rationality and common sense are univocal, despite empirical studies suggesting that there are significant cultural differences revealed in our reasoning capacities.[3] But even if Descartes' descriptive claim about the existence of this sense was shown to be empirically well-grounded in people of all cultures and also convincingly shown to be oriented away from falsity and toward "truth,"[4] there

35

remains a further question as to its value and centrality to the philosophical enterprise. Should this common sense ground and anchor our philosophical methods, or does that commit one's philosophy to theoretical conservatism, to being nothing more than the shuffling of the deck of cards, redistributing things from time to time? But what else might philosophy be? Critique, or "first philosophy"? These are both live possibilities, of course, and a central part of the self-understanding of many continental philosophers, but it is also important to note that the critical disavowal (explicit or otherwise) of the importance of common sense frequently leads to accusations of mysticism and obscurantism, charges that are quite commonly leveled at many of the major continental philosophers. If philosophers are intent to avoid the charge of mysticism, as we ought to be and as virtually all continental philosophers would concur, is our remaining choice between foundationalism (the desire for some indubitable starting point from which other conclusions may be deduced) and seeing philosophy as the coherentist weighing of the balance, the sober judge of what best fits with what, within a given system of knowledge claims, beliefs, and intuitions? This, of course, is roughly the Homeric struggle between the atemporal forms of analytic philosophy and what I have called the more presentist or coherentist forms.

Since the interesting philosophical issue will be to explore what this means metaphilosophically (including in terms of the temporal and normative presuppositions of various methods) and just how powerful an explanatory tool this assessment of philosophy's divided house is, it is worth risking a generalization at the outset. On the question of what role common sense should have in reigning in the possible excesses of our philosophical methods, the continental answer to this question, for the vast majority of the usual suspects associated with this tradition, would be something like "as little as possible," whereas the analytic answer, for the vast majority, would be "a reasonably central one." While this difference at the level of both rhetoric and metaphilosophy is sometimes—perhaps often—problematized by the concrete philosophy of representative philosophers of either tradition, I argue that this norm (and its absence) nonetheless plays an important justificatory role in relation to the use of some rather different methodological practices, thus offering a better account of what is at stake in the divide than many of the alternatives, explaining their respective endorsements of, and resistances to, transcendental reasoning, as well as the paradigmatic way in which work in analytic philosophy proceeds communally.

Of course, it is notoriously difficult to specify any methodological unity at the heart of continental philosophy, which is something of a motley crew, comprised of various different metareflections on philosophical method and with various diverse methods employed in practice: phenomenological, dialectical, hermeneutic, structuralist, psychoanalytic, and transcendental, the last two of which are perhaps most significant in both causing and then legitimating what Friedman refers to as "the parting of the ways."[5] Nonetheless, in different ways all of these methods are designed to shed light upon what might be described as our time-embeddedness, and all are designed to exhibit something that is *not* simultaneously clear and distinct. They are hence sympathetic to varying degrees to Leibniz's (and Bergson's) riposte to Descartes that clarity and distinctness are in fact mutually exclusive,[6] a philosophical objection that begins to explain some of the widely observed stylistic differences between analytic and continental philosophy. In addition, what gives continental philosophy its methodological quasi-unity contra Simon Glendinning,[7] and helps to distinguish it from much of the analytic tradition, is its thorough-going wariness of any close link between philosophical method and common sense (or folk psychology, pre-theoretic opinion, etc.), with the added rider that these reservations are not accompanied by either the strong naturalist or empirical functionalist turn that typifies the majority of those analytic philosophers who dispute the philosophical importance of common sense and conclude that there is no phenomenology, say, or that color perception is illusory. Such a concern is perhaps partly presaged by Kant's critical philosophy (noting that there is a sense in which Kant's transcendental philosophy leaves everything as it is, and is not incompatible with common sense), but it attains renewed vigor around the end of the nineteenth century and in particular with the work of those whom Paul Ricoeur called the "masters of suspicion"—Marx, Nietzsche, and Freud. Nietzsche, for example, famously says in *On the Genealogy of Morals* that, "only that which has no history can be defined," and in *Will To Power* he declares that, "what is needed above all is an absolute skepticism towards all inherited concepts."[8]

Moreover, it seems remiss not to include others in this list, especially Bergson given the challenges his work poses to the categorizing tendencies of the intellect and the simplification that he thinks inevitably results due to its operations, along with Husserl whose phenomenological reduction was designed precisely to suspend the assumptions of the "natural attitude," including even the common sense conviction that we have perceptual experience of an external world. Of course, the relationship between phenomenology and common sense is again not a simple oppositional one, in that phenomenological descriptions typically seek to remind us of prereflective dimensions of experience that have been both forgotten and presupposed. Nonetheless, it is important to note that phenomenology generally aims to describe levels of experience "beneath" the common sense judgments of particular subjects, and which are claimed to be the conditions of possibility for such opinions and judgments. With the poststructuralist thinkers, like Deleuze, Lyotard, and Derrida, this scepticism about any close methodological relationship between philosophy and common sense is heightened.

Of course, it is easy to throw around ideas like critical philosophy as a badge of honor, and the issue of conservatism and radicality in philosophy is more complicated than it might at first glance appear. We all have to pick starting and stopping places in philosophy, and, if we agree there is no one true way, then the question becomes one of just where we draw the line in the sand on issues like the respectability or otherwise of transcendental argumentation, the centrality or otherwise of common sense and its intuitive judgments, and the nature of one's relationship with science. While this chapter is motivated by a general concern to highlight the divergent drawing of the lines that analytic and continental philosophers engage in vis-à-vis common sense, a survey of continental philosophy cannot be attempted here. Instead, Deleuze's direct and sustained attack on both good and common sense will be our prime focus. That is because in its provocative form it provides the most illuminating example of some of the differences that continue to separate analytic and continental philosophy today, and it highlights the temporal component to any reliance on good and common sense. After all, if it is legitimate to maintain that the vast majority of the main figures associated with continental philosophy insist (admittedly to greater and lesser degrees) that philosophical method ought not to be anchored to common sense or other closely related datum, this is not the case with many of the key thinkers and methodological practices of contemporary analytic philosophy. On the contrary, many (perhaps most) analytic philosophers either explicitly invoke the value of common sense, or, as I will suggest in what follows, implicitly value it via techniques like conceptual analysis that want to explicate folk psychology and lay bare what is already embedded in the linguistic norms of a given culture, the widespread use of thought experiments and the way they function as temporally restricted "intuition pumps," as well as the general aim to achieve reflective equilibrium between our intuitions and reflective judgments in epistemology and political philosophy. Such methods enshrine a conservative, or, more positively, a modest understanding of the philosophical project in that it is invested in cohering with both a given body of knowledge and common sense (if not of the folk variety, at least of a given paradigmatic body of experts). After all, Neurath's raft, the favored epistemological metaphor of choice, would be ripped apart by quick and radical revisions to the theoretical architectonic of a given intellectual milieu, and also by any attempt to start philosophy from scratch (i.e., do first philosophy) that did not consider the repercussion for some other domain of knowledge acquisition. The boat can be safely replaced at sea only by taking apart one piece at a time.[9] This accretionist or piecemeal approach to philosophy is not something that most continental philosophers would consider to be part of philosophy's *raison d'être*. While it might be rightly protested that all philosophers are masters of suspicion, wary of *doxa*, clichés, and thus (at least to some extent) wary of common sense,[10] we should note that the suspicion is directed in very different directions: a suspicion of first philosophy and transcendental arguments as opposed to a suspicion of understandings of philosophy as reducible to coherence within a given domain of knowledge claims or valid deductive argumentation.

If this argument regarding philosophical method is borne out in what follows, the foundation will also have been laid for an informed understanding of the limits and possibilities of the methods employed in each tradition. After all, as James Chase and I explore in our *Analytic Versus Continental*, there are particular dangers in an approach that focuses upon (or assumes) this common sense—for instance a bias toward problems of the "right size," and so to necessary and sufficient condition conceptual analysis rather than more diffuse exploration. Likewise, there are different dangers to an approach to philosophy that stridently distances itself from this sense and its associated norms—mysticism, rhetoric, and verbosity, perhaps, but also a research paradigm that is splintered rather than integrated toward common projects. Of course, it is true that continental philosophers have communities and shared justificatory norms within a given community of thinkers to some extent (e.g., existentialism, or phenomenology), and also by association with a particular philosophical name (e.g., Heidegger, Hegel). These shared norms depend upon some kind of valuation of the common sense of a community, and also result in paradigmatic modes of thinking that we might call forms of theoretical conservativism. But my concern here is more with the original philosophers associated with the tradition, rather than the sometimes unproductive exegetical work that can be done in continental philosophy, and while one is always part of a community of some sort, the critical relation to a dead historical community and the comparative lack of attention paid to work within a synchronic dialogic community, means that norms like common sense (and synchronic connectivity with a community of peers and their judgments) are less prevalent in the continental self-image of the job of the philosopher. If this is so, there is reason to suspect that some kind of conversation (even if antagonistic) is necessary for philosophy to avoid some of the weaknesses that can be associated with the methods of both traditions when they become insulated from engagement with their respective philosophical others. If analytic philosophy is methodologically paradigmatic (i.e., it assumes the validity of a shared group of methods that in different ways acknowledge the importance of common sense and the intuitive judgments of a given community), and large parts of contemporary continental philosophy have an a priori rejection of certain kinds of methods on account of their reliance on intuition and refusal to take the transcendental/genealogical turn, without always being able to entirely justify such decisions and preferences, then each needs the other. Debate might rarely yield agreement when the governing paradigm is contested, but it should help to free both traditions of a methodological insularism that can distort theorizing.

Deleuze's Critique of Good and Common Sense

In *The Logic of Sense, Difference and Repetition*, and elsewhere, Deleuze repeatedly discusses two interrelated assumptions that conspire together to produce what he considers to be a false or dogmatic image of thought. These two foundational assumptions are termed good sense and common sense, although we should note that they have a more formal register rather than referring primarily to a given capacity or trait. As he states in chapter 3 of *Difference and Repetition*:

> The implicit presupposition of philosophy may be found in the idea of a common sense as *Cogitatio natura universalis*. On this basis philosophy is able to begin. There is no point in multiplying the declarations of philosophers, from "Everybody has by nature the desire to know" to "Good sense is of all things in the world the most equally distributed" . . . Conceptual philosophical thought has as its implicit presupposition a pre-philosophical and natural image of thought, borrowed from the pure element of common sense . . . We may call this image of thought a dogmatic, orthodox, or moral image (DR 131).

This is a strong claim: conceptual philosophical thought almost inevitably rests on certain unquestioned assumptions, most notably an ideal image of thought (which is also a moral image)[11] that remains tethered to common sense. While Deleuze is not specifically engaging with analytic philosophy in this passage, his repeatedly expressed concerns about any too intimate alliance between philosophy, science, and common sense[12] suggest that he intends this more general description to apply to it. For Deleuze, however, the manner in which this image of thought is betrothed to common sense means that:

> Philosophy is left without the means to realize its project of breaking with *doxa*. No doubt philosophy refuses every particular *doxa*; no doubt it upholds no particular propositions of good sense and common sense . . . Nevertheless, it retains the essential aspect of *doxa*—namely the form; and the essential aspect of common sense—namely the element; and the essential aspect of recognition (DR 134).

Deleuze's view, then, is that much of the history of Western philosophy, as well as its contemporary manifestations in analytic philosophy and phenomenology, has abstracted from the empirical content of *doxa* but implicitly preserved its form (despite the attempt to "bracket" the natural attitude, phenomenology is thought to nonetheless perpetuate an *ur-doxa*). To put it another way, methodological manifestations of good and common sense are the form of *doxa* that persists in various philosophical systems whatever the actual content or conclusions reached. He goes on to suggest, this "form of recognition has never sanctioned anything but the recognizable and the recognized; form will never inspire anything but conformities" (DR 134). Thinking is reduced to recognizing, representing, and to calculative allocation and consistency. Of course, many objections may be raised to this analysis of Deleuze's, not least that it presupposes his conviction that important philosophical problems do not fundamentally require, or even admit of, solutions, but rather call for creative transformations of the problem—hence philosophy is fundamentally about concept creation as he makes most clear (with Félix Guattari) in *What Is Philosophy?* In addition, it presupposes an elaborate metaphysics regarding the importance of difference—as a transcendental condition of the problem, or the Idea—despite the ongoing devaluations it receives. Notwithstanding where one stands on these issues, however, it remains the case that Deleuze's general worries about this dogmatic image of thought are shared, at least to some extent, by continental philosophers as otherwise different as Husserl, Heidegger, Max Horkheimer, Merleau-Ponty, Arendt, Jan Patočka, and Derrida, to name but a few.

What, then, is Deleuze's view of the interrelation and distinction that obtains between good and common sense, and how are they manifested in particular philosophical methods? Common sense is said to be that which allows us to decide on the categories that will be used to determine a solution, as well as the value of those categories. Common sense thus bears directly on methodological issues, including how a problem should be divided up such that a solution might be ascertained. It functions predominantly by recognition (e.g., we recognize that this fits into that category) and is described by Deleuze as "a faculty of identification that brings diversity in general under the form of the same" (LS 77-8). In other words, it identifies, recognizes, and subsumes various diverse singularities (or particularities) and gives them a unity. On his view the notion of a "subject" is a prime example of this process. Good sense then allocates things into the categories, puts things in their rightful place, and selects. It functions by prediction, and by "choosing and preferring," (DR 226) and it is frequently assumed to be naturally oriented to truth. It starts from massive differentiation and then resolves, or synthesizes it. When taken together, Deleuze argues that this model of recognition (including labeling and definitional analyses) and prediction is profoundly conservative. It precludes the advent of the new; good sense and common sense are concerned with the recognition of truths rather than the production of them.

Now, many analytic philosophers might happily accept this designation of their activity as being concerned with the recognition of truths, but Deleuze says that the fundamental role accorded to something like recognition means that common sense finds its objects in the categories of: (1). identity with regard to concepts; (2). opposition with regard to the determination of concepts; (3). analogy with regard to judgment; (4). resemblance with regard to objects. Under the above quadripartite fetters, "difference becomes an object of representation always in relation to a conceived identity, a judged analogy, an imagined opposition or a perceived similitude" (DR 137-8). It is important to be clear about how these "fetters" work. In regard to the issue of the identity of the concept, one example of this would be the tendency to maintain, without a justifying argument, that although a dog has various empirical manifestations (breeds, sizes, shapes, colors, etc.), the concept or Idea of "dog" is nonetheless self-identical. Despite the work of Wittgenstein, Austin, and others, it is arguable that something like this assumption is at play in much conceptual analysis as it explicates our language in terms of necessary and sufficient conditions, and we will see that Stich makes a closely related criticism. In relation to Deleuze's suggestion that the dogmatic image of thought functions through a reliance upon analogical judgment, it is clear that thought experiments are predicated on comparisons between cases and thus on analogical reasoning and judgment. We recognize that a particular thought experiment, for example, is like another (or not) in sufficient respects to stand as a suitable marker for some more general problem of morality, personal identity, etc. This methodological technique hence partly functions by opposition, as Deleuze suggests is typical of this image of thought. Deleuze's fourth claim about this dogmatic image of thought is that perception resembles thought (and *vice versa*), which is another way of saying that it presupposes a harmony between self (or mind) and the world, and this is how Deleuze characterizes the coimbrication of good and common sense in *The Logic of Sense*. In this respect, Deleuze seems to be getting at something closely akin to what Richard Rorty critiques as the idea of philosophy, and indeed the representational conception of the mind generally, as a mirror of nature.[13]

According to Deleuze, these structural features of the dogmatic image of thought apply to us all (both within and outside of philosophy) to greater or lesser extents. Nonetheless, the important question for us will be whether this so-called dogmatic image of thought is not just presupposed by most analytic philosophers, but valorized as a *modus operandi* by some of its most regularly used methods. If the analytic tradition distills and perfects certain methods that are closely associated with this image of thought, and proclaims these features as positive virtues of thought, then we have here a rather stark contrast that calls for elaboration. While it might be protested that Deleuze and analytic philosophers are talking about very different things when they respectively critique or endorse common sense (and associated notions like folk psychology, the philosophical value of our basic intuitions/opinions/pre-theoretic beliefs, etc.), there is a genuine philosophical dispute here. Even if Deleuze is discussing common sense in a formal rather than an anthropological manner (e.g., as some trait humans possess), his argument is certainly that the latter claim still partakes in good and common sense as he defines them.

Common Sense and Theoretical Conservatism in Analytic Methodology

Notwithstanding the optimistic attempts to unify analytic philosophy as concerned with linguistic analysis (Dummett[14]), or reason-giving (Cohen[15]), analytic philosophy is itself typified by various internal fissures. There are, for example, significant divides between the scientifically inclined and those who retain allegiances to linguistic or conceptual analysis, and between foundationalists and pragmatists about justification and truth. Here, however, I want to focus upon two particular methods and two overarching norms that most analytic philosophers subscribe to, even if they do not necessarily put the norms into practice; the methodological use of thought experiments and reflective equilibrium, and the norm of common sense as an important touchstone for philosophical reflection, as well as the associated stylistic norms of simplicity and clarity. These four features are usually interconnected. Analytic appeals to clarity and simplicity underwrite the interest in conceptual analysis, which typically includes not just necessary and sufficient conditions analyses, but also the use of both thought experiments and reflective equilibrium. Although thought experiments often seek to complicate a given conceptual analysis that is purported to be complete (such as Gettier's counter-examples to the definition of knowledge as justified true belief), suggesting that the necessity in question is overstated, it seems to me that Frank Jackson is right to suggest that in philosophy, as opposed to science, most thought experiments are best seen as devices of conceptual analysis, telling us what is conceptually possible or revealing the inferred consequence of a certain definition.[16] The allegiance to an idea of something like common sense (and the assumption of its univocity and orientation to truth) often plays a significant role in justifying the maxim of simplicity, and in according intuitions an important if not constitutive role, whether it be in the use of thought experiments (and judgments of similitude) or the attempt to establish reflective equilibrium. That much analytic philosophy gives a reasonably central role to common sense is perhaps not overly surprising, given that from the outset Moore and Russell allied the emergent movement with common sense,[17] and that others who have been labeled common sense philosophers include some of the greats of the tradition: in no particular order, Roderick Chisholm, A. J. Ayer, J. L. Austin, Gilbert Ryle, Norman Malcolm, Wittgenstein, John Searle, Quine, and even David Lewis.

Of course, these norms and methods do not exhaust the armory of analytic philosophy,[18] and they also quite frequently seem to be abandoned due to considerations from the sciences (e.g., there is no such thing as color). But even the deferential relationship to science that is typical of many analytic philosophers might be brought within the purview of this argument. While common sense (understood either as some capacity that each of us innately has, or the shared opinion of the majority) and science frequently pull in radically different directions, there is another kind of common sense (understood as the opinions of a particular expert community), that is taken very seriously by many scientifically inclined analytic philosophers who would not accept that their projects have any great indebtedness to either of the first two views of common sense. These divergent conceptions of common sense perhaps reflect the Moorean and Russellian inheritances in analytic philosophy and both have a temporal component. In regard to the Russellian inheritance, this would be the supposition firstly that truth (or metaphysical reality) is atemporal, and, second, that the progressivist method of arriving at such truth being synchronic dialogue within a scientific-like community.

On a more general level, simplicity, clarity, modesty, conservatism, and common sense have been envisaged as explanatory virtues in analytic philosophy since Quine influentially gave voice to these norms in *Word and Object*. As Richard Matthews observes, "the primary constraint that we use, qua scientist (for Quine), is considerations of simplicity or conservatism. In Quine's terminology we rely upon the maxim of Minimum Mutilation. We choose to affirm such statement as will minimize the total disturbance to the theoretical system and thereby which best enables a given scientific community to efficiently manage the flux of experience"[19] and philosophy is held to be roughly continuous with science. But such a view is certainly not restricted to Quine. Despite holding a very different attitude on the relationship between philosophy and science, similar views make an appearance at important moments in the work of David Lewis, despite the fact that the invocation of possible worlds is not, at least at first glance, very commonsensical. In *On the Plurality of Worlds*, Lewis says:

> Common sense has no absolute authority in philosophy. It's not that the folk know in their blood what the highfalutin' philosophers may forget. And it's not that common sense speaks with the voice of some infallible faculty of "intuition." It's just that theoretical conservatism is the only sensible policy for theorists of limited powers, who are duly modest about what they accomplish after a fresh start. Part of this conservatism is reluctance to accept

theories that fly in the face of common sense . . . The proper test, I suggest,
is a simple maxim of honesty: never put forward a philosophical theory that
you cannot believe in your least philosophical and most commonsensical
moments.[20]

Despite the first three sentences of this quote (which runs counter to much
of what follows), good sense and common sense, as Deleuze understands
them, are both in evidence in Lewis' comment. Recall that for Deleuze there
are two kinds of judgment: a judgment about categories and their value, and
a judgment that puts things in their place. The moment of good sense comes
when Lewis takes a complex thing, but puts his foot down, and says: just be
honest, tell the truth, be commonsensical, for this is itself an intrinsic good.
Indeed, these terms—honesty, truth, theoretical conservatism, and common
sense—are treated roughly as synonyms here, and the assumption is that
when one is honest with oneself and seriously weighs common sense one is
heading toward philosophical truth.[21] But why think this? Few continental
philosophers would, and it might also be noted that the test Lewis proposes
does not sound like a very rigorous one; we cannot fail to note a certain circularity when a common sense test serves as the basis for adjudicating between
different common sense opinions and intuitions. Nor can this comment be
dismissed as an aberration for Lewis. In *Counterfactuals*, he likewise states:
"One comes to philosophy already endowed with a stock of opinions. It is not
the business of philosophy either to undermine or to justify these pre-existing
opinions, to any great extent, but only to discover ways of expanding them
into an orderly system."[22]

In his defense of conceptual analysis, Frank Jackson proffers a similar
account of explanatory norms, at least in regard to ethics. He suggests: "we
must start from somewhere in current folk morality, otherwise we start from
somewhere unintuitive, and that can hardly be a good place to start."[23] While
one can respect the desire for the philosopher to abandon efforts to attain a
view from nowhere and instead start with certain commonly shared views,
it is nonetheless worth noting the oppositional logic of this formulation that
shuts down a whole range of other possible responses (including "critique"
and "first philosophy") and makes possible a clear and distinct judgment as to
where one ought to start philosophy from—in this case, from the intuitions of
the folk, which seems to serve an equally conservative function to the invocation of common sense or pre-theoretic beliefs. It is also involves a certain
kind of presentism, in that our starting opinions are accorded a significant
role beyond simply being acknowledged as a hermeneutic inevitability. To
recall the quote from Nietzsche that was cited earlier in the chapter, there is
no thought of absolute scepticism toward all inherited beliefs and concepts
and nor is there a real concern with the philosophers of the future as being

somehow different in kind, a prophetic tendency to continental philosophy that Nietzsche was himself a central part of.

On the contrary, for Nicholas Rescher, philosophy's "coherentist methodology requires it to accomplish its question-resolving work with a maximum utilisation of, and a minimum disruption to, the materials that our other cognitive resources provide."[24] If we want rationally defeasible and well-substantiated answers, "this requires that we transact our question-resolving business in a way that is harmonious with and does no damage to our pre-philosophical connections in matters of everyday life affairs and of scientific inquiry." In other words, philosophy ought to harmonize with both common sense and the world, which seems tantamount to arguing that philosophy must mirror nature—nature as revealed by both the sciences and by our common sense, envisaged as a capacity or trait that seems to be an inevitable and largely unchanging part of humanity. Philosophy must build an overall picture of how the various knowledge domains and common sense fit together, and Rescher goes on to say that, "the impetus to economy is an inherent part of intelligent comportment . . . optimisation in what one thinks, does, and values, is the crux of rationality."[25] Without dismissing the value of such philosophical activities, their claim to be the sole appropriate philosophical method can certainly be doubted. As Horkheimer suggests in *Critical Theory*, to take this efficient question resolution as the *telos* of philosophical thought seems to involve a reification of instrumental reason.[26] When reason is understood in Rescher's manner it seems almost inevitable that all that does not fit this definition becomes mysticism, obscurantism, literature, or a related term of rebuke. And, at least in Rescher's hands, such a conception of reason also seems to directly entail a remarkable theoretical conservatism:

> Questions having presuppositions whose truth status is unknown or indeterminate—yet none that are actually (known to be) false may be characterized as problematic. To raise such a question in the prevailing epistemic circumstance is inappropriate because this would be premature in that the question could well become undone by discovering the falsity of such a problematic presupposition.[27]

One need not be a card-carrying Nietzschean to see that this idea of thought seems to be hamstrung by the fear of failure or error, thus forming something akin to an epistemological version of slave morality in which error avoidance is the prime good. While the risk of mysticism may have been assuaged by Rescher's theoretical conservatism, such a manner of philosophizing need not be one that maximizes one's chances of attaining to philosophical truth, just as, given certain conditions, a conservative strategy that focuses on avoiding mistakes will not be propitious in the stock market either.

Now, there are, of course, various points that might be raised in Rescher's defence here, but all of them seem to rely upon what Deleuze called good and common sense, the formal features of the dogmatic image of thought. Common sense, Rescher says, is said to secure his favored methodological principle—Occam's Razor—of trying the simple things first. As he comments: "we subscribe to the inductive presumption in favour of simplicity, uniformity, normality, etc., not because we are convinced that matters always stand on a basis that is simple, uniform, normal, etc.—surely we know no such thing!—but because it is on this basis alone that we can conduct our cognitive business in the most advantageous, the most economical way". He adds that "wherever possible we analogize the present case to other similar ones" and concludes that, "in sum, we favour uniformity, analogy and the other aspects of simplicity because they ease our cognitive labours."[28] It is not difficult to note the similarities between Rescher's inductive presumption in favor of simplicity, uniformity, and normality, and Deleuze's view of the dogmatic image of thought, which persists in the methodological adherance to resemblance, identity, analogy, and so on.

Similar tendencies are also apparent in the work of Pascal Engel, who in an essay on the divide points to a method aligned with common sense and one that lacks it, in a manner not unlike that proposed in this chapter. However, his conclusion that objective cognitive norms are enshrined in the common sense methodology of analytic philosophy, rather than in its continental alternative, is rather hastier than the one I will proffer.[29] He says:

> The standards by which we evaluate our philosophical beliefs should not be different from the standards by which we evaluate our commonsense beliefs. Common sense incorporates implicit norms which go with the very use of such notions as "belief," "knowledge," or "judgment." One of the tasks of philosophy is to assess these norms in an explicit and reflective way, and to evaluate our commonsense beliefs in the light of these norms. The evaluation may lead to revisions of our common sense scheme, and the formation of more sophisticated and theoretical beliefs. But even when we reach these new beliefs, there are no other norms by which we can assess them than those which were implicit in our ordinary practice of forming and evaluating commonsense beliefs.[30]

In such a formulation, Engel allies a process that has come to be called reflective equilibrium very tightly with common sense (we will come back to this), but by highlighting this connection for us Engel also points to a weakness with the method of reflective equilibrium, as well as various other closely related techniques of argumentation. Indeed, as Stich points out in *The Fragmentation of Reason*, analytic epistemology seeks the criteria of cognitive evaluation in the analysis or explication of our ordinary concepts of epistemic evaluation—such as "knowledge," "belief," etc.—which are said to be intuitively accessible to all of us if we care to engage in the required conceptual reflection. But, as Stich insists, this involves a form of theoretical chauvinism, or what Deleuze would call *doxa*, in that our current intuitions about the right standards for cognitive evaluation are socio-culturally produced and to use these to legitimate the given epistemological standards seems to beg the question. Engel, for example, claims that the norms in question come from our understanding of the concepts of "belief" and "judgment" and are said to necessarily hold for cognitive inquiry in any time and place— a form of cognitive monism that he provides no reason for thinking is true. Moreover, on Stich's analysis, and that of other experimental philosophers, the concepts of knowledge and belief seem to be structured rather differently from culture to culture.[31] We will return to this, but for the moment it is time to consider the widespread use of thought experiments, which both solicit our intuitions from us and test those that we already have by comparison and analogy with other cases.

Common Sense and Thought Experiments: The Uses and Abuses of the Intuition Pump

Thought experiments have been part of both philosophy and science for a long time, but they are utilized with greater frequency (and precision) in contemporary analytic philosophy than ever before. Despite the ongoing protestations of figures as diverse as Daniel Dennett, Bernard Williams, and Timothy Williamson, it is difficult to dispute that they are a key methodological feature of contemporary analytic philosophy, constituting a restricted class of the more general concern with counterfactual reasoning. According to Roy Sorensen, thought experiments are the natural test "for the clarificatory practices constituting conceptual analysis . . . a test for which there is no substitute."[32] They aim at clarifying a given position (or concept), often by overturning a given statement by disproving one of their consequences via an expedition to possible worlds. If a consequence of a given proposition might be that a particular situation should not obtain in any possible world (something is ruled out as necessarily impossible), a "necessity-refuting" use of a thought experiment tries to show that the scenario is actually entirely possible/conceivable after all.[33] As such, thought experiments are useful for drawing out inconsistencies in our conceptual distinctions, for prompting us to clarify what we do think and why, and thus also promising to purge us of bias and inconsistency.

Some of the most famous thought experiments have an incredulous science-fiction aspect to them, which is perhaps partly responsible for their controversial status. We might think here of Hilary Putnam's attempt to envisage a brain that is sustained in a vat that would have no need of a body, his twin earth scenario, or Derek Parfit's "teletransporter" scenario, such as it features in his *Reasons and Persons*. As devices of argumentation, these are often useful and enlightening for armchair philosophising: for example, desert island scenarios allow us to bracket away the question of social influences, and the introduction of hypnotists or Robert Nozick's pleasure machine allow the philosopher the freedom to implant certain psychological states into an agent without worries about empirical plausibility. And clearly the production of thought experiments requires imagination and creativity, even if it is arguable that more mundane intuitions show up in the judgments and conclusions drawn from them.

It is notable that they are not, however, regularly deployed in continental philosophy, which instead typically engages with literature and the arts in greater detail, and creativity in this tradition revolves more around concept-creation and phenomenological descriptions. Of course, there are persistent uses of certain stories or fictions that at first glance look like thought experiments, like Hegel's influential "master-slave" dialectic in *Phenomenology of Spirit* (and like Nietzsche's eternal return of difference, or Deleuze's engagement with the Robinson Crusoe tale),[34] but they do not seem to function equivalently. In a certain sense, Hegel's account of the development of self-consciousness via a battle for recognition is a fiction, but it also has another status, being claimed to be both historically evidenced and inferred as a transcendental condition to explain social life and the necessary co-imbrication of the "I-we" relation. Is it short and pithy, and does it allow for a rigorous deductive conclusion to falsify a given statement or conceptual analysis? Is it a test for consistency or best understood as of the form: is x conceivable (or logically possible)? No. It is meant to be both grounded in experiential datum of a phenomenological and historical kind, and, at the same time, world-disclosing, allowing us to look at the world and our place in it anew. It seems clear that what is going on with continental philosophy's preoccupation with genealogical analyses and transcendental arguments (which for Deleuze are often about conditions of possibility of the new and the different: the future) is hence rather different from the a priori concern with possibility of thought experiments.

Although few of the most famous continental figures explicitly explain their reluctance to use thought experiments in any sustained way, their more general reflections on methodological matters suggest they harbor the conviction that something often goes awry in philosophies that uncritically ape the use of thought experiments in science. To put it another way, there is thought to be an intrinsic problem (or at least risk) with thought experiments that stems from the manner in which they strip a problem back to its basics. In other words, the charge would be put that they appear decisive only because of their abbreviated and schematic form, and which might, if filled out, lead to either inconsistency or a failure to discredit a given view. More generally, it is often argued that this abstraction belies the complexity of social life and frequently functions on the basis of certain tacit philosophical presuppositions that are either highly controversial, such as the assumption of a rational, self-interested agent who is extricable from their past, or are merely logically possible rather than practically conceivable.[35]

In distinguishing the phenomenological technique of eidetic analysis (imaginative variation) from the typical analytic use of thought experiments, J. N. Mohanty claims that the latter is based on mere logical possibility, not on what he calls eidetic possibility.[36] Mohanty's claim is that while we may be able to logically conceive of the possibility of, say body-splitting in Derek Parfit's teletransporter device, we cannot concretely imagine this. In other words, it cannot be "lived," or a "concrete intuition" as Husserl might say, although an analytic philosopher might respond, in the manner of Daniel Dennett on a different issue, that this testifies to a failure of the imagination not the general value of thought experiments. But if Mohanty is right, the intuitive responses that a peculiar thought experiment evokes from us are unlikely to be very helpful in soliciting our views about personal identity, or for clarifying and rendering consistent the views that we may already have announced. To put the problem another way, it seems clear that thought experiments need to meet some kind of sufficient resemblance condition to be effective as an argument for or against a given view, and Mohanty's view is that this condition is frequently lacking. What might such a sufficient resemblance condition be? Scientific thought experiments rely on one seeing the resemblance between the imagined scenario and an actual experiment that might be conducted, and ideally will be. In philosophy, of course, things are not so simple, but it seems that one needs to see either a connection to actual experience or to a large body of existing analytic work on the thought experiments in question. Indeed, these scenarios presuppose a shared community of experts for their meaning, familiar with the array of conceptual analyses and the pitfalls that a given thought experiment entails for particular perspectives. As such, they serve as heuristic devices against which views in their neighborhood are sharpened as part of a testing process. The problem with this as a sufficient resemblance condition, however, is that it is rather uncompelling for any philosophers not already enculturated. For those unfamiliar with this background (and not assuming the validity of a given conceptual analysis), thought experiments are hence stale and deprived of depth. For those in the know, on the other hand, the experiments have depth because of the communal work on given problems. That might merely mean that continental philosophers should not dismiss what they do not adequately understand. But is this experience of philosophical depth that accompanies thought experiments genuine or illusory? It can be both.

But it is also worth highlighting that some of these problems are exacerbated when thought experiments are given an explicitly normative or action-guiding flavor. Many thought experiments cut out a time-slice wherein one is asked to imagine a situation without our past or even projected futural possibilities, and to make decisions on the basis of this determination. This highly abstract and "thin" scenario is assumed to nonetheless shed light on our "thicker" practical identity. In Rawls' famous "veil of ignorance" thought experiment in *A Theory of Justice*, for example, we are limited in both of the above temporal ways, having neither knowledge of, or an affective relation to, our past abilities or interests, and having virtually no knowledge of how any futural redistributive arrangements might affect us, and this also applies to the various kinds of thought experiments and rationality paradoxes that one encounters in game and decision theory. Bernard Williams generalizes this point to claim that thought experiments put us in a situation but without our history, including all of the associated information and background that we require in order to make choices. As such, he suggests that they are *exclusively forward thinking*; the past is only relevant in order to predict the future.[37] Does this commit what James Williams calls the "anti-genealogical fallacy,"[38] that is, to ignore the difficult process of tracing the impact of the past within the present (and the future)? Certainly one of the key continental rejoinders to any uncritical reliance on the efficacy of thought experiments would be that they give insufficient attention to the memorial traces (including the unconscious) that constitute (and challenge) the explicit or implicit supposition of a self-interested decision-making agent. If such characterizations can be said to be valid in regard to large classes of thought experiment, it might be objected that when temporally circumscribed in this manner no real decision is possible. After all, any injunction to start from "now," must allow for the phenomena of looking forward to the future (protention and anticipation) and looking back to the past (retention) that constitute any so-called "now."[39] This would be another way, then, of putting the objection that thought experiments tacitly involve a view from nowhere, or what Merleau-Ponty called "high altitude thinking," which ignores the conditions of possibility of a given thought; our always situated background which inevitably involves the past. Given the growing acknowledgment within analytic philosophy of many of these metaphilosophical limitations, however, the key question is whether analytic philosophers are sufficiently cautious in their actual use of such experiments in their work.

In opening up this debate about the philosophical value of thought experiments it is useful to again consider Deleuze. On his view, they remain dogmatic because it is "common sense" that we draw upon to ascertain whether or not the particular thought experiment in question gets a grip on a fundamental moral or political issue, or is an appropriate way of dealing with the dilemma at hand. We recognize that a particular hypothetical scenario—say Plato's ring of Gyges which makes us invisible—stands for a broader problem, in this case the role of fear of consequences in preventing human selfishness (this fear is, of course, removed if we are invisible since crimes could be committed without worries about being caught and punished). A large part of this process depends upon our intuitive response to whether the proposed analogy holds, and whether, for example, Judith Jarvis Thomson is right in claiming that the prolife position that women do not have a right to abortion is analogous to waking up and finding oneself tied to a famous violinist and becoming their effective life support without any possibility of freeing oneself from this arrangement.[40] Usually, the abbreviated form of the experiment or analogy means that the information that is needed in order to make any such adjudication is not given, but we are nonetheless solicited to trust our response. If this is inconsistent with something else we have stated about, say abortion, we are then exhorted to modify our understanding of the relevant moral and epistemological distinctions in order to incorporate that which was revealed by the thought experiment. This is what Deleuze calls good sense, where we attempt to resolve a problem by selecting one set of alternatives over another, or at least providing criteria for such adjudication. The function of good sense is hence to resolve the question at hand by reference to our intuitions on the thought experiment (and its suitability to stand as a marker for the more general problem) and to our rational principles, which we then try to adjust in order to reach reflective equilibrium. In attempting to decide in this manner, however, we run the risk of reducing complex problems to questions which admit of clear and distinct answers, something that might be said of the widely cited "prisoner's dilemma" in which the many different social pressures and desires confronted by two bank robbers who have been caught and are faced with a bargaining situation are simplified into a grid of four possible outcomes.[41] Likewise, Rawls' "veil of ignorance" scenario at times appears to reduce the problem of justice to a judgment between the distributive principles of utilitarianism, liberalism, and strict egalitarianism. In both of these cases problems are understood in a manner that restricts them to a determinate range of possible outcomes; we move from the past as complex and unpredictable to the future as simple, predictable, and amenable to calculation. It is this trajectory of thought, for Deleuze, which gives us the idea of the directionality of the arrow of time, and it assigns to the present a directing role in this orientation (DR 71, LS 88). This is the time of the living-present, as we see explore in more detail in later chapters, although it is a calculative relation to the living-present and hence a particular subset of a richer temporal structure that we will call pragmatic temporality in part 3.

Now it might be the case that this is what most academic philosophy amounts to, but for Deleuze it does not involve a genuine experiment and thus cannot result in genuine concept creation. Why not? Firstly, because the complicated genesis of the intuitions are ignored and the function ascribed to intuition in this temporally circumscribed sense cannot possibly involve a problematization or critique of the socio-political circumstances that have contributed to those intuitions in the first place—as such, it will be bound to the preservation of the status quo. Moreover, it is assumed that this kind of methodological approach to problems—what would you think if confronted with this scenario?—has independent credibility. But should we really trust our responses to an unusual thought experiment, particularly as they concern moral and political life? The problem remains that our intuitions about strange and abstract cases are not likely to be all that reliable, either as a guide to what we really would believe (and do) if confronted with such a scenario, or in regard to what we ought to believe (and do) in concrete situations of any complexity. On the other hand, if we are familiar with the context of a thought experiment, our intuitions about the case in question may well be more forthcoming and perhaps even more reliable, but only in the sense that they agree with what we already knew.

For Deleuze, such a beginning remains pre-critical, within the dogmatic image of thought, where subjective presuppositions continue to wield a pernicious influence despite the immediate appearance of both strangeness and objectivity. Indeed, most thought experiments clearly depend upon common sense as their background. That is, they begin with a certain assumption—for example, we all know that beating animals is bad—and then seek to institute more elaborate examples to tease out precisely what our commitment to this means and entails. For example, is beating animals still bad if they don't feel pain? We begin from a certain common assumption and then recognize that a particular experiment challenges or fits this preexisting assumption. In taking seriously our ordinary intuitions about these kind of non-normal cases, however, we also thereby generalize and extend the orbit and significance of convictions that were, for Deleuze and Guattari, produced in certain particular situations: we subsume the singular under the universal, and thereby privilege reflection and contemplation rather than an affective and intensive relation to the problem at hand.[42] Such a methodological beginning, with its reliance on our immediate intuitions and what "everybody knows," ensures that thought can only orbit around common sense like a moon caught in a gravitational pull. For Deleuze, there is no truly critical philosophy possible from such a beginning, predicated as it is on certain doxic features, including the circumscribed temporal conditions, the reliance upon a generic neutral subject, the validity given to one's pre-theoretic opinions (modeled on the form of recognition), and the manner in which thought experiments are condemned to analogical reasoning. Of course, they can function by revealing a disanalogy between two cases, but in this case it is still a form of analogical thought, one predicated upon bivalence and the institution of an opposition. While analogical thinking can be useful for ensuring consistency, it is no guarantee of attaining philosophical depth. Of course, thought experiments are often just the starting points for analytic philosophers, but if this starting data (and methodological justification) that is fed into a process of reflective equilibrium is partial, then so too will be the result.

Now, it goes without saying that Deleuze is rather tough on both good and common sense (*doxa*), and it is arguable that his mildly utopian conception of philosophy that supports their denigration is itself only one part of philosophy rather than its *raison d'être*. Must we treat the pragmatisms of good sense and common sense (of calculation and intuitive judgments) as harshly as Deleuze does? Good sense and common sense may well predominate in the analytic tradition, explicitly endorsed by many and implicitly enshrined in the methods of thought experimentation and reflective equilibrium, but it might be credibly argued that they are basic and inevitable aspects of thinking. Moreover, it is clear that thought experiments can raise important questions and can even provide the provocation for the institution of a new philosophical problem. The abstraction from concrete situations can have the salutary effect of questioning our everyday assumptions and the rigor of the distinctions that we pre-reflectively draw, but, as Deleuze's work suggests, it also has more pernicious possibilities attached to it, particularly when relied upon as a prescriptive tool, or when taken to answer a question (or exhaust the dimensions of a problem) rather than to pose one. The more difficult and important questions then pertain to matters of degree, which must be addressed in relation to the work of specific analytic philosophers. To what extent does a particular philosopher's use of thought experiments open up areas of inquiry or artificially close them down? But let us move on to the second key method, reflective equilibrium, which often accompanies any use of thought experiments and is also deployed in its own right.

Reflective Equilibrium: Common Sense or Conservatism?

The most famous and influential formulation of reflective equilibrium is found in Rawls' *A Theory of Justice*. Drawing on Nelson Goodman's work,[43] Rawls uses the method of reflective equilibrium in order to explain how we might adjust and perhaps even resolve the difference between our moral/political intuitions about what is fair and just, and the moral/political theory that is endorsed by our rational judgments under the test of the veil of ignorance. But this technique of argumentation is now far from unique to the

Rawlsian liberal tradition. In fact, it has recently been contended that the method of reflective equilibrium is *the* generally accepted methodology in normative ethics, endorsed by many different kinds of philosopher, including both Kantians and utilitarians, and regardless of the philosopher's particular views about metaphysics, epistemology, and philosophy of language.[44] The reflective equilibrium process involves the working back and forward between our provisional judgments about particular cases (intuitions) and applicable rules (principles), with the goal of increasing their coherence, in order to arrive at a more reflectively justifiable—if not necessarily final—position. Goodman's concern in his seminal employment of the method is with what Rawls calls "narrow" reflective equilibrium—a process of coherence adjustment that concentrates on the judgments, rules and background epistemic desiderata that one actually begins with. The intent behind Rawls' own "wide" approach (in *A Theory of Justice*) is to avoid the conservatism inherent in the Goodmanian version of the method. To leave open the possibility of a radical shift in our conception of justice, Rawls suggests that we must bring into play all imaginable sets of roughly coherent judgments, rules and desiderata in some sort of choice situation that is itself governed by such factors, until some sort of stable equilibrium is achieved.[45] To put this another way, there is a difference between an understanding of reflective equilibrium that prioritizes certain starting intuitions or sentiments (which have *independent* value or credibility), and a reflective equilibrium that prioritizes overall coherence with our other beliefs (and hence adds to this equation *dependent* credibility), which might include our theories of personal identity, human flourishing, rationality, the findings of science, and the like.

Now it is clear enough that the narrow version is more conservative than the wide version, since certain of our initial beliefs and sentiments have independent credibility. Assuming their importance in this way arguably makes the narrow method of reflective equilibrium overly invested in the preservation of common sense and what is already (thought to be) known. This is not as clear with the method of wide reflective equilibrium, since it necessarily involves a process of perennial updating, back and forth adjustments between overarching philosophical views, intuitions, and empirical data provided by science. Given that the findings of science are changing all the time, it is clear that an analytic philosopher that uses something like the method of wide reflective equilibrium and gives significant attention to scientific knowledge must consequently also be prepared to change all of the time.[46] As such, an investment in the common sense (of a given expert community) does not seem to be conservative. While this is certainly true, we should also note that often when reflective equilibrium is invoked and claimed to support a given view, no such rigorous process has actually taken place. Time, after all, is finite. While our intuitions may theoretically be subject to revision in wide reflective equilibrium, when it is recognized that we do not have an eternity for decisions and that all of our various beliefs form a part of an intricate system of interconnected cultural convictions, it seems eminently unlikely that the process will involve anything like a radical challenge to our basic convictions and intuitions. After all, even those extra elements that are added to the equation by the method of wide equilibrium (such as consistency with the background theories of science) are themselves at least partly constituted by, and inseparable from, the basic judgments and convictions of the community we are a part of. Moreover, at any particular point in the reflective equilibrium process theoretical coherence is never sufficient on its own but must also cohere with our particular judgments at that time—without this equilibrium our considered position has no justification. The methods of both wide and narrow reflective equilibrium are hence incoherent without according significant value to our everyday intuitions and immediate judgments, and hence, to common sense. Both forms of reflective equilibrium are conservative approaches to conceptual analysis because each is an articulation of a coherentist conception of justification, and radical transitions tend to pull Neurath's raft to pieces.[47]

Although the method of reflective equilibrium clearly bears a structural relationship to the dialectic that has a long history in continental philosophy and reached its apotheosis with Hegel, reflective equilibrium constitutes a particular form of the dialectic. Depending on one's perspective, it is either a dialectic that is anchored in existing social practice by a common sense conservatism that tacitly devalues the new and the different, or it should be acclaimed for this very feature which allows it to avoid the more pernicious forms of relativism (both ethical and epistemological) that can afflict dialectical thinking. Indeed, many of the major continental philosophers would have problems with the strategy of matching our basic intuitions about particulars with a general theory, both because of the conservative implications of the narrow view (which is invested in present opinions), and the manner in which the wide view supposes that through the process of reflective equilibrium we can (and ought to) purge ourselves of any bias that might be betrothed to our initial starting point (or basic intuitions, whether moral or epistemological). This seems to tacitly reinstate another variation of the rational and disinterested subject, and another atemporal ideal of both truth and the subject who is endeavoring to discover the truth, in that it quite radically abstracts from the choices we make and the way most of us live. Indeed, Carl Knight inadvertently expresses this risk when he notes that the subject of any wide reflective equilibrium process "must undergo any experiences that may offset biases in his or her formative influences."[48] This sounds like a noble ambition, but is it tenable, either in the conception of a subject that it presupposes, or in the conception of the philosopher that it advocates? Both the philosopher and the citizen are modeled on the conceptual personae of the judge—as rational, probing, impartial, devoid of affect, and the like. Experiences should be undergone in order to fairly systematize one's thoughts and get them in the broadest possible equilibrium with one another. Now scientists *may* do this in their experiments, but even then something provoked them to think as Deleuze and Guattari emphasize in their accounts of science in *What Is Philosophy?* Moreover, to think that this is how philosophy and a good life ought to be conducted is, to say the least, rather more controversial.

More critically, however, there is a circularity that afflicts the various versions of reflective equilibrium. In *The Fragmentation of Reason*, Stich puts the following objection to Goodman's account of justified inference, and this also applies to Rawls' account:

> What is the relation between rationality and the right test supposed to be, and why is the fact that a system of inference passes some test or other supposed to show that the system is rational? I think the most plausible answer for a Goodmanian to give is that the right test, when we discover it, will be an analysis or explication of our ordinary concept of rationality (or some other common sense concept of epistemic valuation).[49]

In other words, the test case provides necessary and sufficient conditions for rationality because it unpacks our concept of rationality. As Stich points out, however, "for this answer to be defensible our common sense concept of rationality must be univocal."[50] In addition, it must be the case that the procedures used for deciding whether a system is rational must exhaust the concept in question without remainder. Stich says this cannot be a priori supposed and that it is undermined on empirical grounds. But philosophers like Deleuze and Derrida, on the other hand, would seek to contest Goodmanian and Rawlsian claims about reflective equilibrium on something like a priori grounds, and through sustained analyses of particular concepts and their paradoxical logics.

A Role for Common Sense (and Reflective Equilibrium and Thought Experiments), But Not *the* Role?

The point behind this chapter, then, is that from the perspective of Deleuze a methodological reliance on common sense (in offshoots from either Moorean or Russellian forms) entails a theoretical conservatism in analytic philosophy. Now, there are, of course, many possible rejoinders, perhaps most notably, what is bad about being conservative? After all, it is clear that we are and, perhaps should be, conservative in various respects. But the important question for us is whether philosophy *should* be structured conservatively. Arguably it *is* structured conservatively, not least because academia is conservative, but whether it should be structured thus is another question, and one to which Deleuze thinks the answer is no for reasons I have outlined. Another rejoinder might be that there are various problems with Deleuze's philosophy (and I will myself argue this in coming chapters) and with continental philosophy writ large, including its own theoretical conservatism despite the ostensible alignment with critique. It is undoubtedly true that there are myriad methodological concerns that might be posed to continental philosophers about their own insular practices as we will see, but what I want to bring out in this chapter, through the resources that Deleuze's metaphilosophy provides, is an external perspective on analytic philosophy that opens up questions to do with its methodological practice and their temporal significance. It seems to me that the critical views expressed by Deleuze about the pervasiveness of the dogmatic image of thought—and the co-imbricated assumption of good and common sense—serve as timely reminders for the practice of analytic philosophy today, which is methodologically invested in the value of *common* sense (sometimes explicitly, sometimes implicitly), and such a position only makes sense if one assumes that where one is starting from is also *good* (or at least better than the alternatives). In this chapter, a commitment to theoretical conservatism and common sense has been shown to bound up with the particular techniques of thought experimentation and reflective equilibrium. There is a chauvinism bound up with these practices (perhaps not unlike the transcendental arguments frequently deployed in continental philosophy[51]) and they remain questionable and not necessarily inevitable aspects of any serious conception of thought. The uncritical and too prolific use of thought experiments can simplify problems to make them amenable to solution and can lead us to see analogies where there are really significant differences. They might also play a role in leading to a generalizing and categorizing conception of the task of philosophical reflection to the detriment of other possibilities. Likewise, we have seen that the technique of reflective equilibrium is a necessarily coherentist approach that depends on explanatory norms of worthiness that are indeed useful for some kinds of philosophical reflection but not necessarily all. Moreover, there is a circularity to any so-called test of reflective equilibrium: it is not a neutral test, but involves a reinforcement of a given system provided by particular tools found within that system. The problem of conservatism in this practice remains. Another way of putting this might be to say that those engaging in reflective equilibrium almost inevitably find what they are looking for, as in Meno's learning paradox.

But while vigilance about good and common sense and their methodological instantiations is undoubtedly called for, it is not clear to me that we can (or should) understand the genuine philosophical pursuit as ultimately immured of these aspects. Intuitive responses to thought experiments, for example, are undeniably useful for showing us the commitments and contradictions within our own thinking and use of concepts. And concepts are indeed our way of categorizing the world, although we might, as Deleuze exhorts us, still try to create new ones that transform our apprehension of the world. Copernican revolutions in thought have happened, and analytic philosophy should not entirely absolve itself of this hope. The fundamental question then would be: is the new possible for analytic philosophy, given its paradigmatic status and methodological commitment to forms of coherentism and theoretical conservatism? On my view it is possible but improbable, at least without further engagement with its continental (and Asian) "others" on methodological issues, as well as the value of philosophy more generally. This does not entail that such a perspective is bankrupt as a certain reading of Deleuze's analysis might lead us to believe, but nor is a commitment to this kind of theoretical conservatism the only responsible way of proceeding in philosophy as at least some analytic philosophers would have us believe. There are useful roles for common sense, for the coherentist "best fit" approach to philosophy, and for many of analytic philosophy's key associated techniques, but to conceive of them as exhausting the essence of philosophy is to condemn one's philosophy. A philosopher must simultaneously be judge and creator, rather than one or the other.

4

Negotiating the Non-Negotiable: Rawls, Derrida, and the Intertwining of Political Calculation and Ultra-Politics[1]

Arguably John Rawls and Jacques Derrida are equally influential in their respective areas of philosophy and yet the media reception of their deaths in 2002 and 2004 has been drastically different. At least in the Anglo-American context, this involved a formal acknowledgment of the pervasiveness and ongoing significance of Rawls, juxtaposed against a lamentation of the polysyllabic obscurity and ethico-political irresponsibility of Derrida. This emphasis on clarity and accessibility of language, and the curious way in which they have become tools with which the Western media (and Anglophone academia, in relation to the Cambridge controversy in 1993) rebuke Derrida, is through and through political. The assumption seems to be that without simple and unambiguous expression, the values of open communication and free society are threatened. But is clarity and transparency necessarily sociopolitically effective? While Derrida would never simply renounce the ideal of clarity, it is clear that public demands for it can often entail conservative thinking in that they are about reinforcing old values and expectations rather than creating new ones and it goes without saying that genuinely new or revolutionary thinking might not immediately be clear and distinct. Without dwelling further on this question here, which might be glossed as the relationship between style and time, this chapter will examine together the work of these two philosophers, Rawls and Derrida, not least because of the conviction that Derrida (and poststructuralism more generally) offers certain invaluable things to political thought that analytic political philosophy would do well to take account of, particularly as concerns the relation between time and politics.

Now this may not seem immediately apparent. After all, Derrida's emphasis on the radical difference of the future, the "to come," serves as a guardrail against political absolutisms that treat the future as known and

as susceptible of teleological prediction. Derrida's many and varied arguments about the way in which the future disrupts the present, and has its
impact upon the present, without itself being capable of coming to any kind
of definitive presence, precludes this move. Of course, we have seen that
analytic philosophers (including Jonathan Cohen and Nicholas Rescher) have
sometimes aligned analytic philosophy's norm of clarity of argument with the
spirit of democracy, and so it might be felt that this kind of deconstructive
warning regarding ethico-political absolutism has little to say to the analytic
tradition with its clarity and default liberalism. But this is precisely the view
that I want to problematize. Derrida's quasi-transcendental emphasis on the
importance of time and futurity to any understanding of the political is also
useful when employed as a critical tool to examine analytic political philosophy in particular: it highlights that this tradition is often either atemporal in its
calculations, or relies upon references to intuition and common sense in more
or less obvious ways, both tendencies which deserve be subjected to critical
scrutiny for their tacit alignment with a conservatism that wants to preserve
the status quo.

But the argument that I will propose in this chapter is not simply that
figures like Derrida are able to show us some of the presuppositions and problems with analytic political philosophy and with Rawls' work in particular.
On the contrary, although philosophers like Derrida and Deleuze acknowledge the necessity of political calculation, it is also the case that it is vastly
under-thematized in their work. Utilitarianism and liberalism offer two sustained and important attempts at providing such a calculation and it seems to
me that a rapprochement of these traditions is required, fleshing out the kinds
of political calculations that might better respect the significant *moral* insight
at work in poststructuralism. Ultimately I will suggest that phenomenology
and phenomenological conceptions of time are required to expedite this rapprochement, but in order to begin to point to the need for such a political philosophy this chapter will highlights some problems with Rawls and Derrida's
two competing ways of treating the political, juxtaposing Rawls' insistence
upon the calculable and narrower understanding of the political against—or,
more aptly, in apposition with—the Derridean focus upon the incalculable.

Let me try and tease this difference out in a preliminary way. While
Derrida was unjustly vilified for the lack of obvious political significance to
his thought early on (at least compared to his contemporaries like Foucault
and Deleuze), he has more explicitly turned toward such terrain in his recent
considerations of Marxism, refugees, hospitality, etc. It is also the case that he
has always endorsed the more general poststructuralist conviction that philosophical interventions, even artistic and stylistic innovations, are always also
political interventions—a style of politics that might, as Gregg Lambert suggests, be called *une grande politique*.[2] At the same time, it remains the case

that many of the questions that Derrida raises are not through and through "political" in the narrower (and Rawlsian) sense of the term,[3] and this is so according to Derrida's own admission. Instead, he tries to inject a certain thought of unconditionality and incalculability, we might say an "impossible" morality, back in to politics. In his own words, deconstruction can be said to be "ultra-political," or "hyper-political," (R 39)[4] not necessarily in the negative sense that Slavoj Žižek bemoans about an avoidance of the political that afflicts contemporary discourse,[5] but in the sense of looking at the conditions of (im)possibility for the political. It involves a thinking *of* the contours and limits of the political. Rawls, on the other hand, moved in a different direction throughout his career. In his most famous book, *A Theory of Justice*, with which this chapter will be primarily concerned,[6] he not only claimed to have established the distributive principles that must hold for a just society but also to give us clear guidelines as to how they might be implemented— in other words, what kinds of calculations should be engaged in. From this early theory of justice, with all of its Kantian moral-universalist trappings, however, his later "political liberalism" cast aside (purportedly at least) any of his metaphysical suppositions concerning the universal applicability of his theory, as well as the assumption of a particular substantive conception of the self. From a Derridean perspective, Rawls' later work hence seems a particularly ripe target for deconstruction.[7] However, this chapter is more interested in trying to illuminate the necessity of a middle way between an emphasis upon the incalculable that undergirds much of Derrida's work (and his resultant very broad conception of the political) and the much narrower emphasis upon the contingent political calculations that we might make in public in a pluralist society which is Rawls' ongoing focus, as well as his concern with distributive justice which I have suggested, in the introduction to this book, is the main concern of analytic philosophy per se.

Ethics and Politics: The Incalculable and the Calculable in Deconstruction

It must be noted from the outset that there is a tension in Derrida's own *oeuvre* between conceiving of the political very broadly (and such that the acts of philosophy and literature always constitute a political intervention) and conceiving of the political far more narrowly (as pure calculation)—a tension that partly explains his consistently hesitant reflections on whether or not his own work is genuinely political. In "Ethics and Politics Today," for example, Derrida speculates about whether his lengthy interrogations of concepts like the subject, and the human, could be legitimately called political. He ventures that the kind of questioning of oppositions and concepts for which deconstruction has become famous might be said to be "pre-ethical-political." In other words, deconstruction is a condition for responsible political and ethical action. But he goes on to add that this "pre-ethical-political deconstruction" in fact simultaneously *is* ethico-political, precisely because it prescriptively insists on this preliminary "pre" that *must* accompany all responsible decision-making.[8] If we agree with this analysis, we can grant that deconstruction is minimally ethico-political in two senses. First, it holds that there *must* be a trial of undecidability, a moment of madness is necessary for any decision to be more than merely an extension of an existing rule, norm, or law.[9] Second, it contends that there *must* be a recognition of the aporetic manner in which we are pulled in competing directions, for example to be responsible to a singular other, such as a particular loved one, and to also be responsible for all others and the community at large. We can also add a third point that helps to establish deconstruction's political significance. Deconstruction clearly destabilizes existing grounds for political thought and action; grounds that must be overcome if anything new is to appear. It is in this sense that deconstruction can be said to be politically revolutionary and can even be aligned with the spirit of a radicalized Marxism, as Derrida himself claims in *Spectres of Marx.*[10] Moreover, in this text and elsewhere, Derrida insists that the permanent questioning and challenging that is involved in deconstruction (and which opens a space for the new) is synonymous with the notion of democracy itself (Marxism is hence best understood as a radicalization of democracy). Or, to capture his point more accurately, this challenging and questioning of deconstruction is true to the spirit of a democracy "to come"—that is, a democracy that does not make the mistake of thinking that democracy is delimitable to one particular instantiation of it, such as the Westminster system, but is instead about a permanent contestation of the present in the name of an openness to an unknown future.

Although Derrida's work is clearly politically significant under this broader understanding, this is undercut by the narrower and more specific understanding of the political that he himself espouses in multiple places. In "Ethics and Politics Today," for example, Derrida soon goes on to quite radically distinguish the ethical and the political, suggesting that "the most apparent difference between ethics and politics would be a difference of rhythm in their relation to urgency . . . to the interruption of undecidability."[11] He ventures that this rhythmic difference also reflects a difference in structure, and, although provisional, he tells us that ethical responsibility is unconditional and universal in its principles—it is non-negotiable. He suggests that the ethical response should be immediate—it should make straight for the goal all at once, without hypotheticals and calculations. In this sense, ethical responsibility involves a moment of madness, much like Abraham's famous biblical decision to sacrifice his son Isaac that Derrida examines in *The Gift of Death*, and which he and Kierkegaard both read as pointing to the limits of calculative reasoning. Political responsibility, however, is envisaged as calculative and negotiable. It analyses relations of power, involves hypothetical imperatives, and requires strategies and calculated gambles, rather than a mad leap of faith. To sum up, we might say that political responsibility always involves a relative and provisional calculation, whereas ethical responsibility is absolute and incalculable.

As is his want, Derrida then adds something to this formulation, problematizes it, but without wholly abandoning it. He comments that the above is a little too disjunct and suggests that ethical problems occur in the spaces of political calculation, negotiation and deliberation. Politics and ethics are intertwined, mutually dependent upon one another, and, in an important comment, Derrida suggests that:

> What seems not to have to be negotiated politically, not to have to be reinscribed in a relation of powers, thus the non-negotiable, the unconditional is, as unconditional, subject to political transaction: and this political transaction of the unconditional is not an accident, a degeneration, or a last resort; it is prescribed by ethical duty itself . . . One must negotiate the non-negotiable.[12]

In relation to this claim that we must negotiate the non-negotiable, it is worth thinking about terrorism, which governments of all kinds are frequently prone to saying is non-negotiable. Against this view, Derrida points out that saying "no" to terrorism involves a negotiation and a calculation; the negotiation accepts that some singular hostages will die so that terrorism will not be encouraged and ultimately perpetuated. To pretend that this is not so, to refuse to recognize that we all have "dirty hands" and must therefore negotiate in muddy ethico-political terrain, threatens to presage a yet greater violence. Although Derrida begins "Ethics and Politics Today" by representing ethics as the non-negotiable and incalculable, and politics as the art of negotiation and calculative thinking, he hence concludes by suggesting that what is required in the many issues confronting us today (including terrorism) is a means of negotiating the non-negotiable and, by implication, a mutual contamination of politics and ethics, an ability to navigate two "rhythms" as Derrida puts it (with the obvious temporal resonance of this word), without reducing either rhythm to the other.

This kind of argument about the relation between ethics and politics is part of his recent work more generally. In the preface to one of his final books, *Rogues: Two Essays on Reason*, Derrida begins by acknowledging that, "no politics, no ethics, and no law can be, as it were, deduced from this thought [*of deconstruction*]. To be sure, nothing can be done with it . . . But should we then conclude that this thought leaves no trace on what is to be done—for example, in the politics, the ethics, or the law to come?" (R xv). Clearly his reply to this rhetorical question is in the negative, and rightly so. Derrida suggests that his emphasis upon justice "to come," democracy "to come," etc., might in fact be considered to be "ultra-political" (R 39). In this text, Derrida also continues to insist that, "Pure ethics, if there is any, begins with the respectable dignity of the other as the absolute unlike, recognized as non-recognisable . . . Pure politics, which begins with the neighbour as like, or as resembling . . . spells the end or the ruin of such an ethics" (R 60). He goes on to suggest that the political (on this narrow understanding) begins by choosing and preferring the like and that which is knowable, as well as calculable units of measure and axioms. However, it is also important to note that Derrida continually insists that these kind of technical and political measurements—such as calculation and the adding of votes—are not a problem for the incommensurable (i.e., the incalculability of ethics); rather, such techniques are precisely the chance for the incommensurable. There is no incommensurable without, or apart from, the commensurable and the calculable, (R 53) even though the calculation inevitably neutralizes the singularity to which it paradoxically gives access. Similarly, in his now famous essay, "Force of Law," which examines the structurally isomorphic relation between calculable law and the incalculable demands of justice, Derrida insists that justice (and deconstruction) requires us to calculate. Without this, he admits that his own persistent references to justice to come might just serve as an alibi for "the worst."

To summarize these various different texts then, Derrida insists that incalculable ethical absolutes (e.g., justice) need to be put to work in contingent political calculations that are irretrievably context bound (e.g., law). What is needed is a mutual contamination of the political and the ethical that might be termed "ultra-political." I think that this analysis, and in fact the ultra-political emphasis of all of Derrida's work, is exactly right, but some important questions remain. How does Derrida think that we should calculate, accepting his suggestion that we must? There is very little indication of this in his work. Although he regularly insists that there is no pure ethics, no pure justice, any hint as to what kinds of political calculations are better or worse than others is left opaque. Except in the most general terms, Derrida does not engage with the key theories of distributive justice and of political calculations in the narrow sense. While he does discuss the different ways in which we might attempt to add up or calculate equality (according to number, according to merit, etc.), there is little consideration of the relative merits of these different kinds of addition, some of which may be more apt in a contemporary context of globalization (*mondialisation*) than others.

Furthermore, this acknowledged necessity of political calculation is also treated in a more derogatory way than the ethical absolutes that undergird and orient his work. In *Rogues*, for example, Derrida endorses the Husserlian distinction between rigor and exactitude, the latter being more calculative and lesser, and reaffirms that he is on the side of chance and the incalculable because he doesn't want to reduce democracy to a program or procedural system of calculations, (R 132, 5) as we will see Rawls might be accused of. In fact, Derrida also discusses why he generally eschews principles and axioms (or lexical orderings, such as Rawls' famous principles of justice as fairness), suggesting that such principles favor the calculative application of programmatic rules (R 142). It is for this reason that he continues to insist on the "to come," the open-ended and the incalculable, and advocates "postulations" instead of axioms, the distinction being that the former avoids a comparable and calculable scale of values and evaluations.

Moreover, while Derrida regularly uses political terms in his recent work, invoking democracy, the New International, justice, and the like, arguably these terms have become deprived of their content in such a way as to become ethical absolutes. Consider, for example, the notion of democracy to come. Derrida's emphasis upon the democracy that is (and must be) yet to come, means that his vision of democracy is divested of content, calculability, and at least some of its normative force. It also needs to be noted that it is Derrida's general contention that it is precisely that which disrupts calculation and which renders the application of any kind of formula impossible, which is just. For him, justice can never arrive in the present, but is constitutively to come and forever futural—this is another way of saying that justice is non-negotiable and undeconstructible. Such suggestions are part and parcel of his deconstructive practices, which, above all, affirm the new, the messianic, the wholly other, justice, the impossible, and the future, terms that play a closely related structural role in his work. According to Derrida's own definitions of politics and ethics in both *Rogues* and "Ethics and Politics Today," however, these kinds of affirmations constitute an ethical injunction more than a political one, as they are primarily about the non-negotiable and the absolute, notwithstanding that the point of his work is to allow this ethical affirmation to be at work *within* the political.

Interestingly, in an essay titled "The Deconstruction of Actuality" Derrida also acknowledges that his own recent focus upon unconditional hospitality in texts like *Of Hospitality*, and absolute forgiveness in *On Cosmopolitanism and Forgiveness*, is apolitical in a certain sense.[13] His point behind drawing attention to these kind of unconditional ethical horizons that are at work in the concepts of hospitality and forgiveness, is that a politics of hospitality that is purely about calculation and negotiation is a politics that loses all reference to justice. It is merely a machinic system and his point is that we need to ensure that these ideas of *absolute hospitality* and the like must infect the gritty realities of *conditional hospitality*, which is eminently susceptible of political calculation and manipulation. At the same time, it is not enough just to shout, for example, that a given draconian border protection policy is wrong and that a non-negotiable ethics demands that everyone be let in to the country. That is literally impossible on Derrida's view, but what is needed is a softening of the political calculations and a recognition of the importance of this demand of absolute hospitality that sustains and augments any actual occurrences of hospitality, rather than the covering over of that demand. As politically insightful as such a position is, Derrida is reluctant to go any further and specify the kinds of political negotiation of these absolute demands that might be better or worse than other kinds of negotiation.[14] Certain ethico-political stances are ruled out by him (a pure politics of calculation and a pure ethics of the incalculable are both ruled out), but what is left is a wide expanse in the middle within which we must calculate and with which he will not help us much.[15] In that respect, we need to supplement Derrida's valuable but somewhat moralistic insistence that justice is incalculable, for example, with a more detailed examination of the merits and problems of myriad different kinds of political calculations (politics in the narrower sense) that Derrida himself acknowledges are necessary and inevitable. This is where the calculative ambitions of much analytic political philosophy becomes important and useful, whether it be utilitarianism or the liberalism of Rawls that we will now examine, but at the same time Rawls' work also shows us the risks associated with any such turn to calculability—most obviously the threat of engendering a normative moralism and becoming what Derrida would reproach, following Kierkegaard, as a "knight of good conscience."

Rawls: The Veil of Ignorance and Maximin Principle

Unlike Derrida, Rawls has explicitly sought reconstructive political programs and ways in which to best calculate distributions of social goods, and he has sought to find principles to secure democracy's stability (for Derrida, as we will see, it is precisely the inability to find such principles that is notable about democracy and allows him to imbue it with the deconstructive priority evinced by the phrase "*democracy*-to-come"). In *A Theory of Justice*, Rawls insists on the principle of "justice as fairness" and a rational proceduralism about how to bring this about which distances itself from questions of the good life. He argues, among other things, that we should seek to maximize the situation of those worst off in any particular society (often called the maximin principle), and to develop this point he famously asks us to imagine that we are behind a veil of ignorance, unaware of our own station in life. He goes on to describe this original position, where we are forced to strike a fair bargain regarding the distribution of social and economic resources, along the following terms: we are to imagine ourselves in a position of equality in which we do not know most of the socially significant facts about ourselves—race, sex, religion, economic class, social standing, natural abilities, etc. (TJ 12). This thought experiment is supposed to exemplify our intuitive association of justice with fairness, and to tease out its implications. Under this veil of ignorance, we decide what principles we can agree to in order to further our own interests, whatever they may be. Not knowing our position in society, Rawls argues that we are driven by this fact to prioritize those with the worst life prospects and make their situation as bearable as possible, as it may be that we ourselves are the poor or oppressed. Although this is just one of the principles that Rawls thinks that we would choose in this situation, it is important because it offers a tacit justification for liberal capitalism in that social and economic inequalities can be justified if they increase the social goods (such as liberty, opportunity, income) of the least advantaged in a society. Sometimes referred to as the difference principle, it allows unequal abilities to produce differential rewards if, and only if, it is to the benefit of the least fortunate.

Now it may not be immediately obvious that we would in fact choose something like this maximin principle under the veil of ignorance. A Marxist might choose absolute social and economic equality for every person, or perhaps even the distribution of goods purely according to need, as is suggested by Marx and Engels' famous principle, "from each according to ability, to each according to need."[16] Rawls, however, contends that such choices would be irrational if it is granted that allowing certain kinds of inequalities would actually improve the quality of life (understood in terms of social goods) of those worst off, perhaps because of the greater incentives and productivity built into a system that allows for differential wealth accrual. Others respond that under the veil, they might also choose the utilitarian principle of maximising the overall *aggregate* happiness of all people (or even sentient beings), and not worry about fairness and relatively equal levels of happiness between the different people concerned.

It is not wholly clear that Rawls has a satisfactory response to this, at least from within the restrictions imposed by the veil of ignorance. If we are *simply* rational and self-interested behind the veil of ignorance, as he suggests, it seems coherent to take the gamble that we will be happy and rich in a utilitarian distribution of goods, even if the alternative is that we are the slaves who serve to maximize the happiness of all others. Rawls, however, suggests that risking one's welfare being sacrificed for the greater good of everyone else is *prudentially* irrational, because behind the veil we also do not know the probability of whether or not these utilitarian calculations are likely to legitimate some, or many, people being forced into slavery to provide for the overall happiness of the rest of a society. Arguably it is this assumption of prudential rationality which circumscribes the future in a way that Derrida would not endorse (and this is so even when Rawls considers justice for future generations), because in domesticating or trying to protect one's future, the future becomes simply a future present; that is, a restricted and careful relation to the future that is contiguous with all of the expectations and understandings of the present. But given this qualification, Rawls contends that the only rational course of action would be to adopt the maximin strategy, which, as he suggests, is like proceeding on the assumption that your worst enemy will decide what place you occupy in the social system.

Rawls and Calculability

While it is clearly the case that something like what Derrida would call the "unconditional" and the "incalculability" of the ethical has a role in Rawls' work, predicated as it is upon the value of notions like liberty, such features of his work are understood in a restricted sense for at least two interrelated reasons. First, the attention that Rawls gives to liberty is subsumed under considerations of generality, universality, the necessity of a space for public debate, along with a concomitant assumption that what is rational is univocal and shared (TJ 130-5). He accords, on the other hand, little attention to *singular* instances of incalculable demands upon us, and acts of radical freedom, unlike, for example, the unconditionality and incalculability of existentialist conceptions of freedom, which Derrida is sometimes closer to than might be suspected. Second, despite liberty being named as the first of Rawls' lexical principles, considerations to do with freedom and liberty are ultimately subsumed by his procedural preoccupations with fairness, which are of the order of calculability (and, as Derrida's work intimates, the order of calculability entails a synchronic presentism that is not truly open to the difference of the future). In other words, Rawls' interest in human freedom is practically undermined by the general priority that he accords to distributive justice itself. Perhaps Robert Nozick and Peter Singer were right when they suggested that if you begin with a conception of the individual as the bearer of rights (including a right to liberty) it is always going to be very difficult to reconcile this with a genuinely egalitarian social policy.[17] Although Rawls does not go as far as some contemporary decision and game theorists, and although he does not calculate in the utilitarian sense of trying to maximize overall utility, hedons, welfare, or informed preference satisfaction, it is clear that the deductions of the veil and his general philosophical framework endorse a more calculative and narrowly political kind of thinking than that which Derrida engages in.[18] Rawls comes up with a lexical ranking of what is a just procedure and suggests that we consistently need to calculate, insofar as this is possible, whether particular economic inequalities actually do increase the welfare of the poorest in a society. This can be contrasted with Derrida's provocative contention that it is precisely that which disrupts calculation, and that which renders the application of any kind of formula impossible, which is just. This means that justice can never finally arrive in the present, but is constitutively to come, and that in the name of which we contest particular laws, norms, or rules. From this Derridean perspective, however, an obvious question looms: does Rawls' insistence on justice being a calculable state of affairs that can be actualized (or at least that can be imperfectly approximated to) reduce justice to the law and to a proceduralism that might rest content with itself (i.e., have a good conscience), rather than be impelled and motivated by the demands of an incalculable and unconditional justice? In other words, does Rawls' liberalism reduce politics to a narrowly juridical understanding of rights, obligations, and the law? Ultimately I will suggest that it does and that Rawls hence makes the reverse mistake that I have accused Derrida of in his privileging of the incalculable. After all, for Rawls, justice is the state of affairs that obtains when acceptable public rules are honored in decision-making processes, including consequent distributions of goods. This is not surprising, given that it is our ability to choose, rather than the specific goods or ends that we choose, that Rawls considers most deserving of respect (hence the deontology, or Kantianism, of his work). While Derrida also resists subscribing to any simple consequentialism about outcomes,[19] for him abiding by the right procedure in public and stressing the public component to political principles is never going to be sufficient to ensure justice. On the contrary, from the start it denies the demands of secrecy that are at the heart of any relation with a particular singular other (including at the heart of friendship) and it denies the possibility of the event; that which is radically outside of our proceduralist expectations about what justice consists in and is capable of transforming them. To tie this in with another of the key themes of this book, the liberal refusal of the transcendental turn also seems to result in two distinct forms of presentism: first, calculations regarding distributive justice are viewed as what counts in political philosophy (and these typically ignore historical injustice and futural injustice); second, our pre-theoretic intuitions and views must be taken into account, since the aim is to ensure the stability of presently existing democratic societies.

I will come back to this point, but another way of formulating this contrast between these two thinkers might be to say that Rawls does not try to produce a conception of justice, but wants to honor it, or exemplify it. His conception of justice hence loses its utopian element, despite his occasional comments to the contrary.[20] On the other hand, poststructuralist philosophers like Deleuze and Derrida want to produce the new, or, in Deleuze and Guattari's terms, provoke the "people who are missing."[21] Although these two formulations are not reducible to one another, both positions reinstate a minor form of uto-pianism,[22] even if in Derrida's case it is more clearly desertified of content. A political philosophy without this element threatens to be merely that which codifies existing arrangements, the reshuffling of the pack of intuitions and "everybody knows" formulations upon which our social life is, admittedly, partly predicated, but without leaving a space for the new. Although I think that Rawls' calculations are necessary, the question is whether or not in his hands they become reduced to juridical laws and procedures that pay no at-tention to the incalculable. If Derrida affirms the intertwining of the calcu-lable and the incalculable, of negotiating the non-negotiable, but nevertheless ends up prioritizing the incalculable as I have argued, what of Rawls? What role does the incalculable, the unlike, the wholly other (*tout autre*), and jus-tice, understood in the Derridean sense, have in Rawls' work?

Rawls and Ipsocentricity: Does Rawls Immunize Against the Other?

Another way of posing these questions is to ask whether or not Rawls' phi-losophy is "ipsocentric," to borrow a term that Derrida employs in *Rogues* and that connotes self-centric. We have seen that Rawls' philosophy has a conception of reason as the self-interested pursuit of one's own interests (and he specifies that this entails a *prudentialism*), and *A Theory of Justice* also partakes in the famous liberal emphasis upon the freedom of the individual to choose. It might be said that this kind of self-orientation necessarily excludes the other, but such a conclusion is not yet justified. After all, for Derrida, there is a narcissism that is inevitable,[23] and the key issue is whether or not this narcissism that is fundamental to Rawls' theory (and to any other theory) also becomes an egological ipseity: that is, a narcissism that is *immunized* so as to preclude the coming of the other, the different (consider refugees, members of other societies, etc.), and the event.

Immunity is a theme that captured Derrida's attention in his later work. More accurately, he focuses on autoimmunity, taking this term from its bio-logical origins where the self attacks a part of its own defence mechanisms (i.e., immune system), or puts a partial end to itself, in order to live on. It is an ambiguous term in his work, treated somewhat differently in his different texts,[24] but in this context we might summarize Derrida's position as suggest-ing that any self-enclosed narcissism always undermines itself, and that what we both have, and need, is an "autoimmune narcissism" or an "autoimmune ipseity." In other words, we need a rupture in our immune system where our body attacks its own self-protective system and thereby makes possible the event and coming of the other as both a threat and a chance. Interestingly,

Derrida argues that democracy is a political system that is notable for preserving this autoimmune risk. It allows itself to be called into question and it is fundamentally aporetic in that, as has been widely recognized, there are (at least) two competing aspects to democracy: firstly, it is premised upon public voting and the view of the majority, in which we are treated the same as everyone else and are theoretically substitutable for anyone else; but secondly, democracies should also be concerned to respect and look after singularities, or the particular heterogeneities and differences (as well as liberties) of those who do not happen to be in the majority (R 65). In regard to the autoimmune risk that Derrida argues is essential to democracy, it is worth noting that democracies are simultaneously required to leave free those who might make an assault on its defence (immune) systems, but, as we have seen post September 11 in the United States, and also in Algiers in 1992, they might also undermine certain of their own democratic freedoms precisely in order to preserve or save democracy. In the latter case, democracy attacks a part of itself, but it does so in the name of protecting itself, and there is clearly something paradoxical about this process in a democracy, unlike when certain totalitarian governments attempt to safeguard themselves from others and achieve an impossible immunization.

Before we delve more deeply into the question of democracy, it is worth noting that it has been a major communitarian criticism of Rawlsian liberalism that it presupposes a subject who can decide under the veil, despite being divorced from all of their positive attributes and their social situation. Such an account is then criticized for relying on a strange and unargued for conception of personal identity (perhaps an ipsocentric one), as if, at bottom, we are unencumbered individuals divested of all social influences. The communitarians ask: what is left of the self when we exclude our past, our social situation, our vision of the good life, when behind the veil? Could we even really decide anything in this situation at all? This is a part of Michael Sandel's famous and generally perceptive critique of Rawls,[25] as well as one that various Marxists have endorsed. What kinds of replies might the Rawlsian make? They might acknowledge that although Rawls may harbor certain metaphysical suppositions about a personal identity that can be distinguished from the ends that it is oriented toward, it remains the case that we *can* imperfectly conceive of something akin to what it might be like to be under the veil of ignorance that Rawls describes, and, as such, it is a heuristic device to draw out the implications and theoretical consequences of our conception of justice as fairness. In other words, it is not meant to be a theory of personal identity but an expository device that represents equality between human beings as moral people and sums up the meaning of our supposed acquiescence to the principle of impartiality. It must also be reaffirmed that the "original position" is merely one device in his arguments, not the only one, and his notion of a reflective equilibrium still needs to be considered.

Nevertheless, it might be suggested that Rawls' philosophy is ipsocentric if it can be established that it is fundamentally concerned with the preservation of the same, and that this is facilitated by the exclusion of that which is other. In this respect, it is hard to ignore the opening ambit of Rawls' book, when he states that he wants to "formulate a reasonable conception for the basic structure of society conceived for the time being as a closed system isolated from other societies" (TJ 8). This is a strange exclusion, tacitly serving to remain within the paradigm of nation-state sovereignty, and one, it seems, that is not merely accidental to his system. After all, we might also consider his exclusion of animals from moral consideration shortly afterward, which Mary Midgley examines at length.[26] It seems that one of Rawls' key liberal values—conceiving of society as rational and cooperative—ensures the exclusion of the animal, this figure on the periphery, and significant questions remain about whether it is, in fact, the most appropriate premise from which to engage in political philosophy more generally. It is also notable that in part 3 of A Theory of Justice, which seeks to apply the theoretical principles of justice as fairness to some concrete situations, Rawls excludes certain figures—the mad and criminals—from his purview, much as Bonnie Honig has convincingly pointed out.[27] Moreover, the denial of these "inessential" figures is explicitly depoliticized. Indeed, Rawls frequently reassures us that if the principles of justice as fairness were to be fully implemented and institutionalized, then responsibility for dissonances such as criminality would no longer be a communal, societal, or political, but would instead be a personal and psychological problem (TJ 314). He sees no a priori reason to doubt the possibility of justice being actualized without remainder in this way (unlike Derrida), and, in the face of these criminal others, Rawls' theory quickly becomes more exclusionary in its normativity and betrays some of the aspects of his theory that are most responsible for its ongoing appeal.

For example, one of Rawls' key claims against classical liberalism and libertarianism is that having talent, such as a shrewd intelligence, is merely a matter of luck and hence should not be rewarded financially, unless rewarding that talent also happens to maximize the standard of living of those worst off (TJ 18, 73-4). However, in part 3, we are told, by contrast, that criminals must take responsibility for who they are and for what they have done. In other words, early on in his text, talent is said to be contingent and lack of talent is society's misfortune rather than the untalented individual's problem, but later on Rawls holds that criminals are personally to blame for their "nature" (TJ 576), and there is no mention of luck here, despite the fact that one's "nature" would seem to be just as arbitrary as possessing talent. As well as this conflict between Rawls' principles of distributive justice and his practical responses on issues of retributive justice (he claims that the two are inseparable without countenancing the significance of this tension in his work), Honig also points out that there are numerous undesirable consequences of this kind of view that the criminal's "nature" is at fault. Most notably, criminality becomes completely inexplicable, and necessarily so, not just contingently. The criminal is wholly other in the worst sense, radically divorced from the rational consensus,[28] and it seems difficult to see how a criminal could ever by rehabilitated on such a view. They are irresponsible rogues (*voyous*) in much the same way as the United States government has been so quick since the 1990s to designate certain states "rogue states" or cognate terms, as Derrida discusses at length in *Rogues*. They are outside of the principles of justice as fairness and warranting punishment because of this. Even in a state where the principles of justice as fairness are not yet fully institutionalized, for Rawls those who punish in such a system are not sullied by this act.

It almost goes without saying that this is a very different picture to the Derridean account which contends that responsibility is necessarily paradoxical and aporetic—that responsibility to a singular other (such as a loved one) inevitably involves irresponsibility to the other others (the generality) and vice versa. On Derrida's view, we are all hence at least somewhat roguish and this precludes the possibility of having a good conscience about punishment (which would be the ultimate roguery) as Rawls' juridical legislators do. It seems, then, that Rawls tries to immunize his account of justice as fairness against the aporia of responsibility and safeguard his vision of democracy from its others, by excluding those without the right citizenship papers (the *sans-papiers*), animals, criminals, and even the enigmatic and the strange, such as the person whose preference for counting grass rather than doing anything more obviously productive Rawls can scarcely countenance (TJ 432). What becomes apparent is that this attempt to preserve the good from the bad in democracy and political life, and to distinguish the relevant and essential aspects of democracy from the inessential, betrays a conceptual violence, just as much of Derrida's early work has repeatedly insisted this is the case with all forms of metaphysical thinking. Disregarding the validity or otherwise of Michael Sandel's critique of Rawls' metaphysical assumption of an "unencumbered self," we can see that Rawls' attempts to immunize liberal democracy from all that might haunt it remain betrothed to this metaphysical tradition that Derrida describes.

For Derrida, on the other hand, "there is no reliable prophylaxis against the autoimmune" (R 150-1). Although one cannot simply discard prudential considerations, trying to anticipate that which threatens the democratic system and rule it out in advance by immunizing one's self, one's theory, or one's state, against these potential enemies cannot be successful and Derrida argues that this is particularly so with regard to democracy. In the context of his discussion of terrorism in *Rogues*, Derrida insists that rather than it being certain "others" that democratic countries and citizens need to guard against (other countries, foreigners, etc.), this autoimmune risk of the self attacking another part of itself in order to live on is something fundamental to democracy. For him, with the events of September 11, 2001, we have literally seen the way in which the threat comes from within democracy (not from its other) in multiple different senses. To gloss Derrida's analysis, not only were the attacks from within, in that they were both aided (through training) and sometimes perpetrated by United States citizens rather than by members of other nations, but the attacks were also made possible by democratic principles of relatively open borders—democracy must leave free those who might assault its defence systems. In different ways, democracy risked itself, exposed itself, and, as we have seen, the post September 11 reaction of the United States and many countries around the world has obeyed a similarly paradoxical and autoimmune logic—there was a *shutting down and undermining of selected democratic freedoms to ensure the survival of democracy.* Democracy attacks a part of itself, but does so in the name of protecting itself, and this paradoxical situation, Derrida suggests, is constitutive of democracy. Democracy threatens *itself* and denies the possibility of a stable or reflective equilibrium (perhaps even of an overlapping consensus), and this is not something that Rawls seems prepared to countenance. But, as Derrida points out, this autoimmune risk of the self tearing down its barriers is not simply a bad thing. On the contrary, it is the only chance for an exposure to life that is not sanitized against all that might do harm, and it is what ensures that democracy is always an open project, rather than something that could, once and for all, be completed.

In relation to Rawls, Derrida's point would be that he tries to immunize both a certain understanding of democracy, as well as the ambitious theoretical edifice that is *A Theory of Justice*, by denying the possibility of an event (such as a revolution in thinking about what is valuable) because he assumes from the outset that democratic principles are fundamentally good, rational, and self-contained, not riven by this autoimmune logic that Derrida details. Because of this assumption, it also seems to me that Rawls invests his principles of justice with a certain kind of teleology. Even though the interpretation of his key principles and whether or not they have been satisfactorily institutionalized is amenable to some debate, and even though he sets his theory up as explicitly antiteleological (on his view, all teleological theories, such as perfectionism, utilitarianism, hedonism, etc., define the "good" independently of the "right" and the right is then considered to be just that which maximizes the good), the end that his work is oriented toward is the preservation of liberal democracy, and, unlike Derrida, democracy as it is already at least partly known and instantiated, and reliant upon all of our socially grounded intuitions about what fairness consists in, as we will shortly see. In contrast, Derrida accords a priority to a temporality of the future (over habit and the like), largely because he thinks that in political philosophy and contemporary life there is, all too often, a "triumph of the 'it is necessary,' that dispenses with time" (R 109). Democracy and philosophy hence need to give the time that there is not, without resorting to historical teleology's that limit or neutralize the event by imposing a narrative unity that ends up annulling the events of history by over-determining them and undermining the possibility of the new (R 128). While Rawls does not impose a historical narrative upon the events of history, his account is nevertheless teleological in the sense that he holds that it is *only* rational for an individual to live *prudentially* (i.e., carefully) and calculatedly toward the future and to seek to protect (immunize) oneself—otherwise Rawls could not rule out a utilitarian gamble under the veil (TJ 417). As has been pointed out, because of this assumption Rawlsian principles of distributive justice (like the libertarian and utilitarian positions) would not reject the policy of many U.S. states that currently allow people to create private trusts that continue indefinitely, assuming it can be established that it is to the benefit of the least advantaged (if it is not, some minimal redistribution might be called for). Such practices may seem prudentially rational, but arguably it is unjust to permit settlors to benefit their descendants without temporal limitation in this manner.[29] Similarly, on the political and intersubjective level, his major concern is to try to immunize democracy and the principles of justice as fairness against external threats (such as those posed by the future) and ensure its stability. But might not an equally important, albeit dangerous goal, be autoimmunity and the tearing down of the self's protective barriers, the results of which cannot be wholly anticipated? Even if we are not prepared to follow Derrida this far, it is clear that Rawls circumscribes the future and gives it a direction and a bearing that it need not have. His lack of recognition of the resistance of the future to the present, but also of the way this resistance haunts the present and undermines it, becomes a type of teleology in that change is precluded by the conviction that current liberal democratic systems and procedures can be stabilized and are at least roughly contiguous with justice: we might say that, for Rawls, the future is now.

Reflective Equilibrium

Although Derrida and Rawls both privilege justice over other important political concepts like equality and freedom, Rawls insists that the principles of justice as fairness must attain a reflective equilibrium, whereas Derrida contends that justice is always that which disrupts any equilibrium. But what does Rawls mean by this idea of reflective equilibrium? Well, it needs to be noted that Rawls begins *A Theory of Justice* by rejecting what he calls "intuitionism," which is the idea that there are certain principles of action that are self-evident to our moral sense. He suggests that we need a theory that shows why certain situations intuitively elicit disapproval in us, not just that they do illicit disapproval. More famously, Rawls also rejects the focus on rationality of utilitarianism, which engenders all of these well-known situations where we might, for example, owe an unknown Rwandan the same degree of moral concern as we do members of our immediate family. Rawls' notion of a reflective equilibrium is an attempt to avoid this kind of either/or dilemma and to better mesh our rational judgments with our intuitions. As we saw in the previous chapter, Rawls suggests that we move back and forth from our deeply held moral intuitions (common sense) and an analysis of the principles that inform those intuitions, to our considered and rational judgments (TJ 48). While our intuitions may be confused or inconsistent, as they clearly often are, Rawls contends that by advancing principles that accord with most of our intuitions and by reexamining those intuitions that are outside this spectrum, we may move step by step toward a position of reflective equilibrium in which our intuitions are more fully in harmony with our principles. On his view, our considered judgments can and will change as their regulative principles come to light and the veil of ignorance plays an important role in testing our principles. Sometimes we may be forced to modify our account of the original position such that it yields more readily to our intuitions, sometimes we will have to give up our intuitions, perhaps even deeply held ones.

Although Rawls is evasive about the specificities of this process (particularly in *A Theory of Justice*), there seem to be two main structural ways of reaching such an equilibrium. Firstly, by contending that there is a dialectic between our socially grounded intuitions and our considered judgments that is infinitely revisable. On this understanding, Rawls' reflective equilibrium simply negotiates a middle way between intuitionism and the rationality of utilitarianism. Rather than intuitions or rationality being sufficient on their own, we need to play them off against each other until we have a considered and coherent system or worldview. However, any such nonteleological (in the traditional sense) dialectical theory deprives morality of any ultimate ground and becomes a form of relativism as noted in the previous chapter. Given his wavering commitment to moral objectivism it seems that Rawls does not want this, and at other points in *A Theory of Justice* he suggests that there are some initial opinions, or intuitions, that must be accommodated, such as that racism is unjust (TJ 56, 19). If, as it seems, Rawls is committed to arguing that there are some foundational intuitions that get to the heart of the principles of "justice as fairness," then there are many questions that need to be answered that are not, at least in *A Theory of Justice*. Most obviously, he does not justify this claim, nor suggest which intuitions are to be preferred over others and why. Which are the core intuitions that are to be used in this trade-off with our considered judgments? All cannot be, granting that our intuitions are often in conflict with one another. Why even think that our intuitions are helpful at all in cashing out the principles of justice as fairness?

Somewhat dismissively, Rawls suggests that "it would be useless to speculate about these matters here," but he is committed to suggesting that these intuitions are shared, more or less "approximately," between us all (TJ 50).[30] Otherwise, my reflective equilibrium is likely to be seriously divergent to the reflective equilibrium of Pope Benedict, and we are no closer to ensuring the stability of the principles of justice as fairness and indeed democratic systems themselves. It seems that the only way a reflective equilibrium at a communal level could ever be reached is if most of us agreed on those fundamental intuitions, at least to an extent. But if everyone agrees in the first place, then there is little to argue about and the process of coming to a reflective equilibrium is merely the systematization of the pack of intuitions that we always already shared.[31] While Rawls' might accept this designation without seeing its critical force (after all, his theory of justice is meant to exemplify social practice), such a reduction in scope of political philosophy raises some difficult empirical questions. Arguably we do not have such intuitive agreement and Rawls simply presupposes that we share similar intuitions about what fairness consists in: it is a liberal notion of fairness that ignores consideration of both desert and merit.[32] But do we have intuitive consensus on this exclusion of merit? What about hard work, for example? Do we agree with Rawls that it is not a relevant factor in the fair distribution of social goods?[33] Is his understanding of fairness the only, or even the best conception of fairness? Derrida would certainly problematize this assumption, perhaps even by reinforcing the claims of, say, a meritocracy upon our notions of fairness.

But there is also a more fundamental problem with this reliance upon intuitions in Rawls' work. From various different continental perspectives, seeking such a match between theory and moral intuitions obscures the question of the genesis of the intuitions in question, and forgets about the unconscious and the possibility of various Marxian notions that have been influential even in passing out of fashion, including false-consciousness and ideology. While he acknowledges that ordinary moral thought and intuitions may sometimes be deeply divided, even contradictory, the unconscious has no place in Rawls' analysis, and this, it seems to me, is highly significant. Even if the metaphysical reification of the concept is frequently unhelpful, it seems clear from both psychoanalysis and the poststructuralist appropriation of it that we must not disassociate questions of desire and pleasure (which are often not fully conscious) from the political. In fact, if there is consensus on anything in continental philosophy since Husserl, it would be that a

pure epistemology is impossible, and that what is required is instead theoretical adumbrations of the interconnections between knowledge, power and
desire. Foucault and Deleuze and Guattari are enjoined in this respect. The
latter pairing contend, for example, that "the question of desire's involve-
ment in its own involuntary servitude is the fundamental problem of political
philosophy."[34] Moreover, our unconscious desires, which are at least partly
determinative of our intuitions, are, as Freud has famously and I think quite
plausibly told us, polymorphously perverse. They are multiple and in conflict
with one another and will also vary widely between each of us. If this is so,
what kind of consensus on this intuitive level might there be? If psychoanalysis is broadly true, or more minimally if it is conceded to teach us something
of enduring import about subjectivity, it has significant implications for any
philosophical modus operandi that privileges intuition and the rational subject. Intensities and affects do, after all, derive much of their force from non-
present times—that which has been, or that which might be—and we explore
this in relation to Deleuze's work in chapter 6.

To momentarily extend our purview beyond Rawl's work in order to
elaborate on this worry, consider Brad Hooker's rule-utilitarianism in *Ideal
Code: Real World*. While Hooker's view doesn't postulate absolute consensus about people internalizing societies rules, a certain level of acquiescence
is nonetheless required for the rules he is advocating to be functional and
efficacious. Hooker suggests that this means "internalization by the over-
whelming majority, everywhere, in each new generation," and he also settles
on a figure: 90 percent.[35] Of the 90 percent who internalize the given rules
that are thought to maximize well-being not all are thought to comply, but
Hooker nevertheless presupposes that it will result in "general" compliance.
This claim seems contentious, resting on his tacit conception of the subject
as rational and self-interested. Against this position, continental philosophers
would typically maintain that psychoanalysis shows us that internalizing a
rule does not entail complying with it (consider, for example, the phenomena of sadism and masochism that we will return to). On the contrary, what
becomes clear is that there will always be forms of resistance, conscious or
otherwise—we might even suggest that laws and rules create resistance and
noncompliance—and Hooker, from this perspective, does not spend enough
time exploring the gap between internalized rules and behavior. This "gap" or
"between" is something that continental philosophers continually thematize
in different ways. In fact, for them it is the condition of possibility of ethics
(of responsibility), and, once more, any theory predicated on the absence of
such a gap is anethical, merely about the judicial application of rules and law.

Rawls on Rationality and the Lack of Undecidability: Derrida on the Autoimmunity of Rationality

Rawls' avowed view of rationality, at least in *A Theory of Justice*, is simply the efficient and self-interested pursuit of one's ends; that is, instrumental rationality. In his descriptions of the veil, for example, he implies that the sole reason that people cooperate and socialize is for their individual advantage,[36] and this view of rationality fits neatly with some of Derrida's descriptions of politics as about calculation and negotiating prudentially to achieve a desired outcome. Of course, Derrida's whole point is that there is another aspect to rationality, what might be called an *aprudential* and ethical element: that is, the incalculable demands of justice, of absolute hospitality and impossible forgiveness. Without these dual and simultaneous demands, there is no aporia, and, for Derrida at least, there is also no decision and no political responsibility. Do Rawls' characterizations of the veil of ignorance have anything to say about this? After all, as has been frequently observed, it might be said that there is not actually anything like a fair bargain taking place in this situation as the possibility of substantive differences has been precluded from the beginning by the details of the original position. Indeed, it is important to note that while we are asked to wager that we might be poor, or lack intelligence, we are not asked to consider that we might be criminals, rogues, or even a Bodhisattva who does not value Rawls' social goods of liberty, opportunity, and income, or the basic assumptions of his vision of prudential rationality.[37] It seems fair to say that there is a strong normativity embedded in Rawls' account of the veil; it is about fixing and orienting our thoughts on what is just, not on challenging them in order to create new conceptions of justice. Indeed, despite Rawls' conviction that what most deserves respect and moral consideration in humans is our capacity to choose (rather than the particular ends or visions of the good life that we do choose), there is a strange and inconsistent denial of choice that goes on in his work that is particularly in evidence in his descriptions of the "veil." As Honig observes, it functions as an anaesthetic that gets rid of the possibility of any diversity, any contest, and therefore any choice.[38] After all, in the original position we are all essentially the same. Since noone knows what position they occupy in the social system, asking people to decide what is best for themselves (self-interest) is the same as asking people to decide what is best for everyone considered impartially (benevolence) and the possibility of genuine conflict is ruled out on. The good of others becomes subsumed as part of our own good (despite the apparent liberal insistence on individuality and free choice), much as is also the case with utilitarianism. In fact, this emphasis upon generality arguably governs most analytic considerations of responsibility and justice. Responsible behavior toward others is about establishing general principles, publicizing them, and then deciding or calculating in accord with them. This is certainly Rawls' view, when in *A Theory of Justice* he details some of the formal constraints of his concept of right as opposed to the good, and specifically emphasizes generality, universality, publicity and what he calls "finality," suggesting that there is no further court of appeal, so to speak, to determine the rightness or wrongness of an action or policy (TJ 131-5). For Derrida, however, to exclusively emphasize this aspect of responsibility toward others (generality, calculability, and substitutability, rather than singularity and incalculability) is to ignore the constitutive paradox, or tension, which allows us to speak of responsibility at all.[39] There needs to be a moment of undecidability, an aporia, a tension, and a leap of faith; otherwise there is no responsibility of any sort, political or ethical. This is missing from Rawls' descriptions of the veil, in which we are all ostensibly the same and the choices made behind the veil are hence strictly deductive (TJ 121).

Of course, it is important to note that Rawls' use of the veil of ignorance scenario is not committed to saying that this is, in fact, how "real empirical selves" decide (against Sandel's interpretation). On Rawls' account we are not required to wholly rely on the veil, but should also consider our common sense intuitions and attempt to negotiate a reflective equilibrium between them. The key problem with this, I think, is that both the rational (understood as calculative and prudential) deductions of the veil, and the intuitive reliance upon common sense, are, on their own, inadequate paradigms for political reflection. Most notably, the picture of rationality itself is insufficient, only half the story, in that the demands of unconditionality and incalculability are either set aside, or are domesticated and deprived of their futural orientation as mere intuitions and common sense. Indeed, it is this univocal understanding of rationality which explains Rawls' juridical-like ease of calculation. He is certainly convinced that dilemmas are not undecidable, whether it is inside the deductions of the veil, or faced with real life bargaining situations. For Rawls, there is a right answer. Even in his later work, which focuses on the reasonable rather than the rational, and where what he calls "reasonable comprehensive doctrines" are not susceptible to any kind of definitive adjudication, it remains the case that conflicts between liberty and equality are settled once and for all,[40] whereas for Derrida their relationship remains inherently aporetic.[41] In part 3 of *A Theory of Justice*, Rawls even suggests that it is conceivable that there be a culture without resistances to it, a state without power,[42] and, at a later stage in this text when he imagines a society without the need to punish, it is not because it is magnanimous but because no one disputes these just and fair institutions that reflect the maximin principle. There is, as William Connolly suggests, an "ontology of concord"[43] assumed at the basis of his work; Rawls exhibits both the desire for, and assumption of the possibility of, a harmony without remainder or excess. In the process, he downplays the inevitable violence at the heart of normalization. Indeed, while Rawls is theoretically heavily in favor of equality and the welfare-state, and for this he is to be commended, the spectre of Marx (which Derrida insists upon the importance of) is noticeably absent from his work, precisely because of this assumption of concord and because there is little acknowledgment of both normative and structural violence (e.g., the violence at the heart of capitalism and wage-labor transactions). Violence, risk, and chance, are repressed as Rawls presents us with a situation where there is no fundamental tension and refuses to recognize the inevitability of dirty hands. It seems to me that political thinking needs to recognize the impossibility of this Rawlsian harmony, of stable and teleological *reflective* equilibriums, which concern judgment, opinion, and overlapping consensus,[44] rather than the embodied movement toward an equilibrium that I will argue in part 3 of this book is unjustly ignored.

As we have seen, Derrida troubles this kind of analysis by pointing to the autoimmunity at the heart of democracy, by highlighting the necessary roguery of all state sovereignties, as well as the aporia at the heart of any attempt to be responsible. Perhaps even more significantly, *Rogues* shows us the dual and competing demands of rationality. In an important passage toward the end of the book, Derrida suggests that we have to "think together two figures of rationality that, on either side of a limit, at once call for and exceed one another" (R 149). To think both the heterogeneity and the inseparability of all these pairs of oppositions that we have considered thus far—calculability and incalculability, politics and ethics, law and justice, equality and freedom—is to bear witness, Derrida argues, to an "auto-delimitation that divides reason." (R 150) In other words, there is an autoimmunity at the heart of reason itself, and Derrida insists that both calculation and the unconditional are necessary. In theory at least, he advocates a "perilous transaction" between these polarities and suggests that, "what is reasonable is the reasoned and considered wager of a transaction between these two apparently irreconcilable exigencies of reason, between calculation and the incalculable . . . between human rights and unconditional justice" (R 151). Discussing the particularities of human rights and calculating which modes of distributive justice are more or less likely than others to ensure equality is necessary, but we also need to continue to deconstruct in the name of the incalculable and in the name of justice. Of course, we have seen that Derrida and deconstruction tend to side with the incalculable, with ethics, and even with a freedom that is not linked to sovereignty and autonomy (which is on the side of the incalculable) over equality (which is on the side of the calculable). But perhaps there is also a need (in relation to political philosophy) to reverse the practice of deconstruction without betraying its spirit; that is, we need to side with political calculations while keeping a vigilant eye on the mad, the incalculable, and all that is susceptible to, and in need of, deconstruction. While Rawls' early work has shown us some of the considerable theoretical risks of a too narrow focus upon political calculations (ipsocentricity, normative exclusions, a conservative reliance upon common sense and intuitions, etc.), it is my wager that we need to bring this kind of calculative focus to bear with, and upon, both deconstruction and poststructuralism more generally. This would clearly involve a more pessimistic and interrogative form of calculation; a calculation that acknowledges the violence of state sovereignty and of democracy,[45] rather than the good conscience understanding of them of Rawls, the legislator, who seeks our reconciliation with them.

Part Two

Poststructuralism, Time Out of Joint, and Future Politics

5

The Politics of Futurity in
Derrida and Deleuze[1]

In all of their work Derrida and Deleuze aggrandize the importance of a futural dimension of time, although it must be noted from the outset that we cannot understand that to entail a valorization of the future as a particular place, or empirical moment in time, from which one will be able to judge the trials and tribulations of history. That kind of relation to the future (the future understood teleologically, as a "soon to be present") runs counter to their attempts to emphasize a conception of the future whose import lies in the way in which it simultaneously interrupts any "now" or present moment and is also a condition for that living-present. Moreover, they consistently link their valorization of the future with a theorization of the event (a key aspect of the work of virtually all of the poststructuralists), and assert that it is this aspect of the radical singularity of any event that needs to be emphasized as a counterbalance to conceptions of time in which the future is treated as known and demarcated according to either the expectations of the present (habitual), or the predictions and calculations of the present (reflective). Moreover, much of the political thrust of their work derives from elaborations upon this point: there is an ethico-political impetus accorded to this interruption to the temporal order that opens upon the unknown, and that cannot be treated as simply a future that will one day become present.

This apparent proximity between them is interesting, not least because there was relatively little explicit textual interaction between them over the course of their long careers. Derrida's eulogy for Deleuze is powerful and evocative (see *Work of Mourning*), and they often express respect for one another, but there is not much philosophical interaction between them that merits the name.[2] This might seem to attest to some important philosophical differences between them and Gordon Bearn has hence suggested that, "the difference between Derrida and Deleuze is simple and deep: it is the difference between No and Yes."[3] Daniel W. Smith offers a more subtle variation on this argument when he suggests (with some reason) that Derrida is a philosopher of "transcendence in immanence" (and for a Deleuzian this means that the figure of negativity has returned), whereas Deleuze is a philosopher of immanence.[4] Likewise, Deleuze proclaims that he is still a metaphysician, whereas deconstruction operates on metaphysics but is not simply another form of it. Perhaps, also, it might be coherently argued that Derrida rejects or at least troubles empiricism, whereas Deleuze endorses a new form of it in his conception of transcendental empiricism (DR 56, 144, 147). The status of Deleuze's work in this regard, however, is exceedingly complicated, poised between a pre-Kantian philosophy of immanence that is indebted to that "prince" or "Christ" of philosophers, Spinoza, and a post-Kantian philosophy of time and the transcendental.

Despite such stark differences, in many important respects Derrida and Deleuze come together on the question of time in general, and more specifically on the priority that is to be accorded to the future. Despite Derrida's famous early suggestion that there is no concept of time other than the metaphysical one, as well as his repeated suggestions that the play of *différance* itself (with the temporal deferral that it implicates in all things) can never really escape the realm of representation and therefore metaphysics, Derrida's recent work increasingly emphasizes a conception of futural time that cannot be integrated within the horizons of a subject, and he makes these kind of temporal points in relation to the structure that he terms the "messianic," as well as his somewhat synonymous emphasis on that which is "to come" in relation to democracy and justice. Do Derrida and Deleuze refer to similar things when they talk of the future in this way? If so, does this mean that Derrida's philosophy abandons, or at least challenges, his earlier formulation regarding time necessarily being metaphysical? After all, if this radical future is that which breaks open time, then it seems that it cannot be subsumed under metaphysical conceptions of time. It becomes more like a *quasi*-transcendental condition for time, and this, according to my understanding, also brings Derrida closer to the work of Deleuze. It is to these questions and their political consequences that this chapter will return, but for now our task is a simple one—to explicate the broad outlines of Derrida and Deleuze's often shared future politics, as well as to intimate how this fundamental position will be developed, complicated, and sometimes argued against, in the rest of this book.

Derrida and the Time of the Future

There are many themes from Derrida's recent work that unremittingly emphasize the future. In *Politics of Friendship*, to cite one notable example, he argues that the thought of the perhaps is the only responsible thought of the event and of the future. In fact, Derrida declares that the "perhaps" is the only just category of the future (PF 29), and this is because it respects the unknown rather than circumscribing it in any particular direction, thus tying in with Deleuze's repeated insistence that problems and questions are more important than solutions and answers.

Democracy to Come

In a lot of his later work and particularly in *Politics of Friendship*, *Rogues*, and *Spectres of Marx*, Derrida has frequently discussed the idea of a democracy "to come." His point in this regard is relatively simple. A genuine democracy is supposed to be about facilitating difference and facilitating the exposure to multiple perspectives and outlooks. But if democracy is the name for this openness to what is to come, then there is a paradoxical sense in which democracy itself can never actually arrive—democracy is impossible, at least in the peculiar sense that Derrida imbues the notion of impossibility with (PF 22). After all, it seems clear enough that to declare that democracy has finally arrived, once and for all, in a particular contingent historical format is contrary to the spirit of democracy. Similarly, to suggest that a particular government instantiates democracy, even an altogether admirable kind of government, is contrary to the spirit of democracy; it is to perpetuate a kind of fascism in that it closes off the future, and involves shutting one's eyes to that which might, and will, force us to reevaluate. Democracy, then, is impossible, because it is about an openness to the future (difference) that cannot become present.

But Derrida would also want to claim that the notion of a democracy-to-come cannot simply function as a regulative ideal either, because we cannot know exactly what kind of difference the future will bring with it. There is not, Derrida suggests, an ideal form of democracy, comprised of a list of finite characteristics, that we aspire to, and can get closer and closer to. Any such regulative ideal treats the future as partially given, in that there is an expectation that our current aspirations will be contiguous with our future aspirations. In *Rogues*, he details three major objections to the surprisingly frequent interpretation of his work as tacitly positing a regulative idea: (1). any such ideal tends to remain teleological, of the order of an ideal possible that is infinitely deferred; (2). responsibility cannot consist merely in following, applying, or carrying out a norm or rule; (3). and for unspecified reasons internal to subscribing to the role of the regulative ideal in the broader Kantian project (R 83-5). In sum, Derrida resists associating his understanding of the future with a regulative idea because it would be to domesticate genuine futurity and genuine otherness. It would be tantamount to the subjective anticipation of the future that, as we will see, Deleuze argues is involved in both habitual and memorial time. Finally, Derrida also resists this kind of regulative ideal because of the ethico-political exclusions that it can help to legitimate, ensuring that there is no place at the table for the wholly other, the foreigner, or whatever resists our current (and hence finite) understandings of democracy and its subjects. Although a responsible democracy is never unambiguously assumable, there is a suggestion that a more responsible government would be open to the future in this way, open to the democracy to come.

Justice to Come

Similar themes are raised in Derrida's many discussions of justice, which he also argues must be constitutively "to come." In "Force of Law," Derrida emphasizes that every deconstruction is undertaken in the name of something that is undeconstructible—justice, openness to difference, the wholly other, the marginalized, the "to come," or the future.[5] These are all roughly synonymous, and Derrida makes clear that without the undeconstructible, deconstruction would be without motivation. But what motivates deconstruction is neither something specific, nor a vague utopian outline of an ideal world. On the contrary, it is motivated by something that is "desertified" and abstracted of all concrete content: something that is unforeseeable. We might say simply that deconstruction is motivated by the future, understood as the pure and empty form of time. But to return to the issue at hand, justice then is also to come, deferred, and can never finally arrive, as was the case with democracy, but this does not commit Derrida to an ineffectual pessimism. Rather, it means that every time the law, or a rule of conduct, folds in upon itself and becomes overly legalistic, such as a government responding in a populist way to their electorate and initiating a "three strikes and you are out" policy (three breaks of the law and you are incarcerated regardless of circumstance), then the law needs deconstruction in order to give justice a chance. The point is that justice and the law require each other, and further, that the one without the other is useless. Justice and the law are not supposed to be opposites, but to interweave with one another; as John Caputo puts it, "laws ought to just, otherwise they are monsters, justice requires the force of the law, otherwise it is a wimp."[6]

In order to understand this mutual implication of justice and the law, it is helpful to think about Derrida's suggestion that a judge needs to invent the law, for otherwise we could simply get a computer to dispense judicial punishments, paying no attention to the singularity of the participants and the event of transgression. In an important sense, a judge needs to exceed the letter of the law in order to be just, and this is the case even where mercy provisions are themselves enshrined in law, because those particular supplements to the law are also permanently susceptible to revision.[7] At the same time, it remains the case that a judge cannot completely ignore the law, but must negotiate this tension and must, as Derrida enigmatically suggests, "negotiate the unnegotiable."[8] It should be clear from this that to deconstruct does not mean to destroy. Rather, to deconstruct the law (and Derrida means the law in its most general sense) is to open it up to alternative meanings, such that revision is possible, and such that there can be a space for justice. There is no end to this deconstruction, however, and justice does not arrive. Justice cannot finally arrive, but is instead the radical future that haunts the time of the present. The logic of this "hauntology," as Derrida puns in *Spectres of Marx*, means that justice is impossible in an important sense. One cannot claim that this or that social organization is just, as justice is not a present thing. It is an openness toward difference (the future) that is always both betrothed to the law, while simultaneously interrupting the calculations of the law, and undermining any attempt to compulsively stick with the law, or to take any social organization as self-legitimated. Again, the implication is that a judicial or governmental policy, in order to be just (insofar as this is possible) needs to keep the future open and permanently revisable.

The Messianic

In order to understand what Derrida is getting at with these various invocations of the "to come," it is also worth explicating a term that Derrida has borrowed from Walter Benjamin and the Judaic tradition more generally. That term is the messianic and it relies upon a distinction with messianisms. According to Derrida, the term messianism refers predominantly to the religions of the messiahs—i.e., the Muslim, Judaic, and Christian religions—although he also suggests that there are philosophical messianisms, as we will see. These religions proffer a Messiah of known characteristics, as described in the respective religious texts and oral traditions, and often one who is expected to arrive at a particular time and place. The important point is that the coming messiah of the future is known, and *his* characteristics—maleness is almost inevitably one such characteristic—are foreseen. In an obvious sense, this futural messiah is hence not acknowledged as wholly other, or beyond human comprehension, but rather is anticipated as conforming to certain spatio-temporal characteristics according to the (phallogocentric) expectations of the present. Now, Derrida is not simplistically disparaging religion and the messianisms they propound (John Caputo, Kevin Hart, and others, have illustrated this at length),[9] but Derrida's desertification of the messianisms puts a decidedly different spin on things. His call to the wholly other "to come," is not a call for a fixed or identifiable other of known characteristics, as is arguably the case in the average religious experience. On the contrary, his wholly other (*tout autre*) is indeterminable and can never actually arrive.

Derrida more than once recounts a story of Maurice Blanchot's where the messiah was actually at the gates to a city, disguised in rags. After some time, the messiah was finally recognized by a beggar, but the beggar could think of nothing more relevant to ask than: "when will you come?"[10] Derrida intimates that were the messiah actually to turn up, this would be a disaster. He suggests it would shut down the structure of time and history. It would close off the future, as well as the hope and faith that this radical temporal difference makes possible. For Derrida, the messiah must hence be constitutively to come, or if they have already come, they must be to come again, one final time, and there are many religions that do suggest this, Christ and Christianity being the example par excellence. Even when the messiah is claimed to be present, there is a sense in which he or she must still be yet to come, and this temporal necessity reaffirms the force of the distinction between the messianic and the various historical messianisms that Derrida details in *Spectres of Marx*. Whereas the historical messianisms domesticate the unknown and the future, the messianic refers to a structure of our existence that involves openness toward a future that can never be circumscribed by the horizons of significance that we inevitably bring to bear upon that possible future. In other words, Derrida is not referring to a future that will one day become present (or a particular conception of the savior who will arrive), but to an openness toward an unknown futurity that cannot be anticipated. It is time desertified of any concrete content and is hence formal, and deconstruction is clearly motivated by a desire to open up myriad possibilities for a future that is of a fundamentally different order to the "now."

In order to avoid the problems that such messianisms can engender—killing in the name of progress, mutilating on account of knowing the will of God better than others, etc.—Derrida suggests that: "I am careful to say 'let it come' because if the other is precisely what is not invented, the initiative or deconstructive inventiveness can consist only in opening, in uncloseting, in destabilizing foreclusionary structures, so as to allow for the passage toward the other."[11] This means that Derrida cannot and will not name the future. For him, there is only a persistent hope for what the future might contain and a deconstruction that entertained any type of grand prophetic narrative, like a Marxist story about the movement of history toward a predetermined future that would, once attained, make notions like history and progress obsolete, would be susceptible to deconstruction. For Derrida, any dogmatic Marxism that is not primarily about an immanent critique of capitalism quickly becomes a philosophical messianism, because the future is determined in a particular way. While more will need to be said about all of these complicated issues in the remainder of the book, we can bluntly summarize: Derrida wants to emphasize the way in which the future is not, in fact, foreseeable, but is radically other (even though it must inherit from the past), much as Deleuze's *Difference and Repetition* also does, albeit in a different idiom that lacks such an *obvious* political engagement with questions concerning justice and democracy. Of course, Deleuze's more obvious political engagement comes in his collaborative works with Guattari. While not wanting to undermine the significance of these highly influential works, it seems to me, however, that they derive much of their political impetus from Deleuze's two key philosophies of difference, *Difference and Repetition* and *The Logic of Sense*. If the key political dimension of the collaborative works with Guattari consists in contrasting a molar politics of equality and identity to a molecular politics of becoming and difference, we will soon see that this analysis is heavily indebted to the transcendental (and causal) privilege that Deleuze accords to the time of the future (the eternal return of difference) and an incumbent affirmation of multiplicity without identity.

Deleuze and *Difference and Repetition*

Deleuze's most systematic examination of time comes in *Difference and Repetition*, although *Bergsonism* and *Cinema 2: The Time Image* accord the temporal sustained attention and *The Logic of Sense* offers a particularly nuanced explication of his view of the relation between the event and the pure form of time that he there calls *Aion* and counterposes with the time of *Chronos* (I engage with that material in the next chapter). As is now well-known, in *Difference and Repetition*, he attempts to establish, contrary to much of the philosophical tradition, that repetition is, and never could be, the simple repetition of the same. Instead, he argues that all repetition involves the instantiation of difference, although not necessarily to the same extent, or in the same way. A key part of this project is his demarcation of three different approaches to time, all of which, he contends, involve repetition, as well as difference in repetition, although it must be noted that it is only the final synthesis of time that achieves the reversed Platonism that is his ambition (DR 59): that is, to understand *repetition for itself* and *difference in itself* (DR 94). His three temporal paradigms are habitual time, memorial time, and futural time, although these should not be understood in the traditional common sense way as simply being three different empirical components of time. Rather, they are three representative ways of looking at time as a totality, and Deleuze suggests that they are hence exemplified in the work of certain philosophers, but rather than being a purely transcendental description he also argues that these are concrete structures of experience that can, in a certain sense, be lived. Habitual time is the time of the "living-present" in which the synthesis of the chain of present moments constitutes time. Memorial time is the time of the past, as events are placed into unitary order of time (e.g., in reminiscence) that imposes a form upon sensory experience. Finally, there is the time of the future, which is the eternal return of difference. Deleuze argues that all three of these modalities involve a synthesis of time (with active and passive varieties), and the implication of this is that we cannot have an experience without synthesis, although this is not a Hegelian synthesis

that admits of sublation. In order to grasp what Deleuze's association of the
futural synthesis of time with the eternal return of difference is referring to,
and why it is ultimately prioritized (both in a transcendental/ontological[12]
and existential/moral sense), it is necessary to begin by explicating the two
other fundamental modes of time that Deleuze describes in *Difference and
Repetition.*

First Synthesis of Time: Habit

As mentioned, for Deleuze, the first synthesis of time is that of habit, which
gives us the phenomenological experience of the living-present. On this view,
time is constituted by an originary synthesis that operates on the repetition
of instants; it contracts the independent instants into one another to constitute the living-present. The past and the future then become but aspects or
dimensions of this living-present (which is not itself an instant): the future is
that which is anticipated to occur, whereas the past is the preceding instants
and background conditions that are retained in the contraction that makes up
the present (we will shortly see this illustrated in Deleuze's consideration
of the recurring AB AB AB AB series). This living-present also sets up a
directionality or arrow of time in that it goes from the past to the future, and
from the particular to the general. In that sense, the habitual synthesis of time
is aligned with what Deleuze calls and denigrates as common sense, which
goes from most differentiated to least differentiated, and, as Deleuze says,
also provides for the arrow of time. Questions remain about whether it is fair
to say that common sense and habit (understood as a lived skill) are equivalent and also about what kind of politics can afford to pay so little positive
attention to these phenomena. We will return to this, but we must note that
for Deleuze habit concerns, "not only the sensory-motor habits that we have
(psychologically), but also, before these, the primary habits that we are" (DR
74). On Deleuze's partly Humean view then, habit is hence the condition of
the self or ego that accompanies the contractions of the living-present. It is
worth interrupting this explication here to suggest, with a view to what is to
come in this book, that both Derrida and Deleuze aim to sideline the importance of habit and embodied coping (perhaps especially when it comes to
the ethico-political realm), and we can also note in passing that this thereby
entails a rejection of any politics that privileges the "I" or the self, as many
forms of liberalism have done. For Deleuze, the "I" is produced by myriad
habitual syntheses of time, so to begin with it is to begin with a contingent
and derivative phenomena (and this also applies to issues like human rights).
For him, individuality may be said to be universal but subjectivity is the dam-
aging illusion of universality (DR 258).[13]

But without getting too far ahead of ourselves, it is clear that habit involves some kind of repetition. When we behave habitually, we repeat previous modes of conduct, admittedly within certain loose and quite flexible parameters, such as trying to park at the university in places where we have previously found a parking spot. It is a more difficult problem, however, to establish how habitual behavior might also involve difference, as Deleuze insists it must—his fundamental contention is that difference and repetition are not simple opposites, but that each is necessarily embedded in the other. For Deleuze, habit also involves difference primarily because "habit draws something new from repetition—namely difference (in the first instance understood as generality)" (DR 73). Habit is hence not simply a mechanical repetition. Rather, it also involves a prereflective recognition (based on the passive synthesis) that the activity that is being engaged in is something that has been done before, and this is the generality of which Deleuze speaks. For example, in the famous Humean series AB AB AB AB, it is habit that introduces a difference between one set of the series and the next. It is habit that creates a difference between the two repetitions and leads us to expect a B whenever we encounter an A. As Deleuze suggests, "when A appears, we expect B with a force corresponding to the qualitative impressions of all the contracted ABs. This is by no means a memory, nor . . . a matter of reflection" (DR 70). His point is that it is not that we reflect upon the past, or even consciously remember the past—in fact, Deleuze argues that repetition induces us to repress any conscious recognition of that repetition (DR 93-110)—but that we simply know how to go on in a nonreflective way. This passive synthesis of time occurs *in* the mind prior to memory and reflection, but it is not carried out *by* the mind. While this process partially depends upon the past experiences that are involved in the synthesis (and what is often, perhaps problematically, called procedural memory), but at least according to Deleuze, there is no memory involved in the living present of habit. In other words, he contends that it is not because of memory that the second pair of the group AB AB differs phenomenologically from the first.

The main premise of a recent film directed by Christopher Nolan, *Memento*, offers a good example of the manner in which habit might be envisaged as being largely independent of memory. To briefly summarize the story, the main character, Leonard Shelby, witnesses his wife being raped and perhaps murdered (this is ambiguous), and thereafter suffers a memory loss problem. Lenny cannot create new memories after the incident. He still remembers incidents prior to the rape, but after that everything passes and cannot be recalled. Every minute or two, it is an abyssal difference that returns; everything is new, repeatedly. For example, in one scene Lenny has no memorial retention of how he got in the shower, but suddenly just finds himself there. For our purposes, the important point is that Lenny does not remember his past at all, and yet he can still develop habitual responses toward his situation that ensure that he can learn and develop increasingly refined ways of adjusting to the world. How is this possible? By simply repeating particular actions, aspects of his behavior become quasi-instinctual. He begins to habitually know where to look for things. Even though he has no recollection of what he did a minute ago, it is nevertheless possible to learn and to adapt to what is ostensibly an absolutely new situation, through the deployment of consistent patterns of behavior. This is something that Merleau-Ponty also consistently emphasizes in *Phenomenology of Perception*, where he insists upon the epistemological importance of the way that the body acclimatizes to its environment, and thereby makes possible an embodied learning and training, and, in fact, the vastly different significance that they accord to these phenomena will concern much of the second half of this book.

Second Synthesis of Time: Memory

While this habitual explanation of time, in which the chain of events or passing present moments constitute time, seems to offer an adequate explanation of the constitution of the living-present, the present also passes in the time that is thus constituted (it can be exhausted and is hence not coextensive with time per se), and his basic question is hence something like, "why is it that a habitual present, or temporal 'now' moment, can pass", or, "why is it that the present is not totally coextensive with time?" Deleuze suggests that this necessarily refers us to a virtual or transcendental condition for the living present, as Bergson also maintained. To put this another way, there needs to be a *second* synthesis of time that causes the present to pass, and this, he argues, is the time of the past, or memorial time (DR 79). The fundamental idea is that we cannot represent a former present (i.e., the past) without also making the present itself represented in that very representation. So, if we think about our past, we also in some sense bracket away the present, or cause the present to cease to be. This means that whenever we remember, there will be two main aspects to this: firstly, the "actual" memory of that past; but also a representation of the present (or the self) as itself being engaged in remembering. Deleuze describes these two aspects as memory and understanding. According to Deleuze, then, this second synthesis of time, the past, is the *ground* that means that any present always necessarily passes (as it is bracketed away in the attempt to remember), and it hence allows for the arrival of another present (DR 80-2).

Again, *Memento* serves as a helpful tool in clarifying how this memorial aspect of our lives might cause the habitual present to pass. After all, what Lenny cannot experience is memory and this means that Lenny's living-present of habit does *not* pass in the ordinary sense. Rather, there is only an abrupt break, or gap, in the temporal order, and Lenny suddenly and repeatedly finds himself in new situations. Deleuze argues that memory involves a more profound synthesis than habit, and he makes a very Derridean point to disrupt any privilege that might be accorded to the living-present of habit. He argues that "no present would ever pass were it not past 'at the same time' as it is present; no past would ever be constituted unless it were first constituted 'at the same time' as it was present" (DR 81). So there is no temporal "now" moment, and the present must itself always be divided. The present is always already differentiated from itself, and we have seen the manner in which the repetition of habit does, in fact, instantiate a difference (it introduces a difference between the first AB and the following AB, by causing us to expect a B whenever we encounter an A). For Deleuze (as for Derrida in his discussions of the trace), any invocations of the past refer to a "pure past" that has never yet been present. As Jon Roffe has put it, memory "synthesizes from passing moments a form in-itself of things which never existed before that operation."[14] In other words, unity is imposed upon the past in that it makes of the past this or that specific meaning. Difference, or pure difference, is hence denied by this synthesis, but it is also clear that Deleuze does not want to hold that memory is a simple matter of recollecting or re-presenting the past. Memory also creates the past, and hence instantiates a lesser form of difference.

From the perspective of this conception of time, "the past, far from being a dimension of time, is the synthesis of all time of which the present and the future are only dimensions" (DR 82), and it involves a very different conception of both the past and future than those which were at play in the first synthesis. With memorial time, the past becomes the reflected and mediated past of representation and of reproduced particularity. In this respect, it is helpful to consider the phenomenon of reminiscence and contrast it with the absence of a past of this kind in the ready-to-hand pragmatics of locating a car-park or observing the AB AB AB AB series. Whereas habit contracts different instants from the past with a view to anticipating what is to come, memory contracts a differential level of the whole (DR 84). The former is material, the other spiritual, or, as Deleuze says at one point, virtual. In memorial time, the future also ceases to be the immediate future of habitual anticipation, but instead becomes the reflexive future of prediction based on the generality of the understanding that memorial time has made possible. While understanding is built on the habitual synthesis of time, it transforms the relation to the future from one of embodied and pragmatic anticipation to one of rational calculation, and the dispute between these levels will concern us in our attempted rapprochement between phenomenology and analytic philosophy. For the moment it perhaps suffices to say that political struggle works at many levels, even if Anglo-American political philosophy might not always reflect this, and we will consider embodiment and the living-present in more depth in part 3.

Deleuze also goes on to make the provocative claim that "every reminiscence, whether of a town, or a woman, is erotic" (DR 85). Although this is never precisely explicated, his point seems to be that the experience of time in this second sense is structurally akin to the experience of desire. In this memorial dimension of time, we violate the purity of an absolute past and fail to understand it in its difference and singularity. Likewise, desire is an experience in which we never completely get what we want, so to speak, but must always leave intact the mystery or singularity of that which is desired. However, Deleuze has another reason why he considers the temporal modes of habit and memory to be insufficient, and this is explained in some detail. Basically, he argues that these two conceptions of time do not properly institute time in thought. Although we no doubt *can* think habitually, and *can* regurgitate prior ways of thinking about problems (but remember that for Deleuze, habit must always be flexible and nonmechanistic), this is not genuine thought. Similarly, although there is understanding involved in memory (as the present is also bracketed away with the past, and this enables the necessary distance for something tantamount to reflection) this is also not genuine thought, because it depends upon the traditional understanding of representation and reason that he refers to as the dogmatic image of thought. As has been discussed ad infinitum, for Deleuze there are four principal aspects to "reason" insofar as it is the medium of representation: identity, in the form of the undetermined concept; analogy, in the relation between ultimately determinable concepts; opposition, in the relation between determinations within concepts; resemblance, in the determined object of the concept itself. As we saw in chapter 3, these are the "four shackles of mediation" and, for him, resemblance, identity, analogy, and opposition are effects, the products of a more primary difference. Deleuze argues that truly philosophical thought involves a futural form that breaks open time, and interrupts time, even if it always also pertains to time.

Third Synthesis of Time: Futurity and the Eternal Return

Both Derrida and Deleuze refer to Hamlet's declaration that time is out of joint and link it to the pure or empty form of time that they associate with the future, and which is tightly linked to, if not synonymous with, the event. In fact, Deleuze defines this order of time "as the purely formal distribution of the unequal in the function of a caesura" (DR 89). However, Deleuze (unlike Derrida to my knowledge) quickly associates this affirmation of pure difference, of the future, perhaps surprisingly, with Nietzsche's famous thought of the eternal return of the same. While Nietzsche regularly affirmed the importance of his conception of the eternal return, arguably the most important recent treatment of the concept can be found in Deleuze's work, although Lyotard and Vattimo also discuss it. Deleuze accords the eternal return sustained attention in several different texts including *Difference and Repetition*, *Nietzsche and Philosophy*, *Pure Immanence*, and his essay "Nomad Thought." Basically, he endorses Nietzsche's suggestion that the thought of the eternal return can be an aide to a revaluation of values, but he also transforms it by putting a peculiar inflexion on Nietzsche's notion of the eternal *return of the same*, instead exalting the eternal *return of difference*. He suggests that it is only the eternal return of *difference* that represents the time of the future, rather than history repeating itself in the return of the same, and he also argues that this is at least partially congruent with Nietzsche's own stated position.

While Nietzsche suggests competing and sometimes mutually exclusive interpretations of the eternal return, at least according to the traditional interpretation, the eternal return poses a single but complicated question that can be schematized as follows—what if a demon informed you that whatever mode of action you choose now, it will be repeated indefinitely? The implication of this is that everything recurs as we have once experienced it, and an infinite number of times. This view also has potential cosmological implications if we are to concretely attempt to conceive of it, but Nietzsche seems to intend the eternal return to function more as a test designed to intensify experience, and to ensure that we choose to do something that we are prepared to affirm over and over again. The presumption seems to be that such an individual, who can affirm her own life on the supposition that her chosen action will recur infinitely, can handle anything. This conception of the eternal return seems to represent the world, or at least an individual's life, as repeating itself endlessly in identical cycles. Nietzsche explicitly suggests that a life is envisaged as returning in sequential order, and that there will be "nothing new in it."[15] It would hardly seem to be a good candidate for thematizing the future.

According to Deleuze, however, to emphasize this one conception of the eternal return in Nietzsche's work is to misconstrue the main import of the idea, which he argues concerns the return of *difference*. Now there are undoubtedly some intuitive problems with the notion of difference returning— how can difference return if it does not share something with that which has preceded it? Is it even coherent to claim that difference can return? However, it is precisely this kind of logic that Deleuze wants to reject in his insistence on understanding pure difference, and he instead argues that the eternal return is designed to affirm only the contingent nature of things, and the manner in which we are confronted with a certain situation and compelled to act. Like Nietzsche, there is a sense in which Deleuze also wants to affirm the eternal return as a thought that is designed to test us. However, he argues that any commitment to the eternal return affirms only that initial experience of contingency—the moment of not having any particular mode of action necessitated, or of openness to the future. Whereas the Kantian categorical imperative prescribes a certain course of action (at least if one is being rational), the Deleuzian/Nietzschean test affirms the way in which the thought of the return does not prescribe anything, and it hence affirms chance, and that things may have been other than they are.

Elsewhere, Deleuze also goes on to argue that what is affirmed in any genuine thought of the eternal return is not the whole, as the Ancients may have held, and as some of Nietzsche's comments also indicate. Rather, Deleuze argues that the eternal return is selective (DR 91-2), and that it frees us from conventional morality. We choose to affirm something in particular (which rules out and *differs* from any number of alternatives). To invoke an example of Deleuze's, if one were to choose cowardice or laziness to recur eternally, there is a sense in which this kind of affirmation of these attributes would actually transform them to such an extent that they could not be adequately called cowardice or laziness any longer.[16] So the eternal return is designed to intensify and affirm our experiences of each moment, and Deleuze argues that it is the affirmation itself that recurs eternally, and not the consequences, or even the agent who does the affirming. Reaction, negation, and sameness do not return, and the consequences of actions do not return. It might be protested that if this is the case then it seems difficult to use the Deleuzian conception of the eternal return as a test to intensify experience. If the agent does not return, then why worry about what you affirm, or why imbue your decision with any intensity at all? In the Nietzschean test, the motivation to do so is provided by the possibility that actual consequences, and an agent who suffers them, will recur indefinitely. In other words, there is a question here as to how the Deleuzian inversion of Nietzsche's position might function as an aide to a revaluation of values. It rejects traditional normative morality (much as the return of the same also does), but the return of difference, in which agency plays no part, and in which the will also gets left behind, does not seem to motivate or intensify action in quite the same way. Deleuze and Nietzsche seem to be speaking on different levels. Even though Deleuze insists on the importance of the existential experience of the eternal return, to a greater extent than Nietzsche he is also naming conditions for the possibility of experience (hence his more ontological conception of the eternal return), as well as conditions for that particular kind of experience that Nietzsche is trying to encourage. This brings Deleuze into closer proximity with the work of Derrida. Rather than juxtaposing the Nietzschean Deleuze with the more transcendentally concerned Derrida, it seems that the issue is more complicated than that.

What is required, for the moment, is to connect this discussion of the eternal return of difference with Deleuze's descriptions of the final temporal mode—the future. Deleuze frequently connects these two ideas, but what could such a conjunction mean? Clearly it suggests that for him, the eternal return of difference is more than merely a thought experiment, and it is not simply about history repeating itself. Rather, for him the eternal return functions as a paradigm case of the futural dimension of time that he ultimately wants to privilege. Among other things, we can take this apparently paradoxical insistence upon the return of difference as a critique of Hegel, in which difference is continually reabsorbed by the synthesizing movement of the dialectic. For Deleuze, the thought of contradiction is not yet the thought of genuine difference (DR 263), and this means that thinking must not construe difference as derivative—or to be overcome, as he argues is the case with Hegel—but rather as fundamental to the possibility of any relation at all. Of course, a complication for Deleuze's position is that he argues that this third temporal mode remains a synthesis of time. The obvious response to this is to ask what exactly is being synthesized in the return of difference? If difference can be synthesized, then it would not seem to be an experience of pure difference, precisely because the contradiction, or difference, is overcome in the synthesis. In other words, how can one synthesize pure difference? For Deleuze, such questions are badly posed, as throughout *Difference and Repetition* he argues that the relation (the difference) precedes the actual terms or polarities of the synthesis. The synthesis that is involved in futural time, and the experience thereof, is hence not recuperative, but instead must be conceived of as disjunctive, in that the only unity of the eternal return is the negative unity that difference does, in fact, return. He repeats this claim when he suggests that, "it is not the same or the similar that returns, but the eternal return which is the only 'same' and the only resemblance of that which returns" (DR 126).

To put this third temporal modality into more experiential terms, an experience of the future, the time of the eternal return, would hence be oriented toward change. Of course, that does not tell us much, but Deleuze argues that he cannot tell us much about the eternal return, as pure difference cannot be adequately represented in conceptual thought.

In order to try and intimate what is involved in difference-in-itself (i.e., the pure time of the future) without subordinating it to the regime of representation, as well as to shed more light upon the way in which this idea might contribute to a revaluation of values, Deleuze resorts to images and narratives of characters (or what in *What Is Philosophy?* he calls conceptual personae) confronting particular events. He discusses the past, or memorial time, as involving a disposition in which the act, or the event (whether it be imagined or empirical), is envisaged as too big for one. He suggests that this attitude is often exemplified in the cases of Hamlet and Oedipus. But these two figures can also sometimes be taken to stand in as markers for the habitual time of the present, in which Deleuze suggests that there is a becoming-equal to the act (marked by Hamlet's sea voyage and Oedipus' enquiry respectively): the hero becomes capable of the act and equal to whatever event may transpire and, with a view to our forthcoming consideration of the work of Dreyfus, it might be said that they have mastered a technique. However, Deleuze tells us that in the futural synthesis of time, any such acting self is abandoned, as is the embodied comportment and equilibrium of the agent coming to terms with the act. In the time of the return of difference (pure time, the future), "the event and the act possess a secret coherence which excludes that of the self" (DR 89). This is a closely related point to Deleuze's ethical injunctions to become worthy of the event (or, in reverse, not to be unworthy of what happens to us) in *The Logic of Sense*, which are likewise associated with the exceeding of the self and the time of the present (Chronos) by the time of Aion as we will see shortly. In this context, one might also recall Deleuze and Guattari's comment that: "when something occurs, the self that awaited it is already dead, or the one that would await it has not yet arrived."[17]

But why exactly must a genuine thought, or experience, of the future be subjectless? We can see from Deleuze's Humean-inspired understanding that the subject is the product both of a habitual synthesis, and also a retrospective imposition in order to give past disparate moments some kind of overall meaning or unity. These memorial and habitual syntheses set up the basic conditions for subjectivity, but subjectivity is not part of the time of the return, the time of the future. It is for this reason that he suggests that Nietzsche's leading idea is to ground the repetition of the eternal return in both the death of god and the death of self (DR 11). For him, if the future is to genuinely be the future, then it must not be restricted by this kind of identity. Rather, the future is pure difference, or pure temporality, without the identity of subjectivity betrothed to it, and the "esoteric truth" of the idea of the eternal return of difference hence concerns the idea that the eternal return affects only the new, the unanticipatable, or the future as such, and not specific agents or conditions which return. Subjectivity anticipates the future, projects toward the future, and thereby deprives the future of its genuine futurity—it makes of the future a future-present. Again, this is not a genuine exposure to difference, but is a domestication of difference and the future. As Deleuze suggests, the time of the eternal return not only changes the grounding aspect of memory into a simple default condition, but it changes "the foundation of habit into a failure of habit, a metamorphosis of the agent" (DR 94)—it is this latter claim, and the privilege associated with the failure of habit, that will concern us throughout this book.

In this third synthesis of time, Deleuze enigmatically suggests that, "the order of time has broken the circle of the same and arranged time in a series only in order to reform a circle of the Other at the end of the series" (DR 91). Now it is this circle of the Other that Deleuze describes, as well as this tacit claim that pure difference can be experienced, that Derrida does not so clearly endorse. For Deleuze, it is only the yet-to-come that returns, and he wants to argue that there is a sense in which one can experience the return of difference (as well as it being a transcendental condition for this temporal triumverate that cannot itself be thought representationally) without that difference being subsumed by identity. This isn't quite the case with the later work of Derrida, despite it becoming both more Levinasian in its insistence upon the Other and also more Deleuzian in its insistence upon the future. But what is the status of Deleuze's claim that the eternal return of difference (the future) is both the totality of the temporal triumverate and also its end? How might this kind of ontological claim aide any revaluation of values? How might it allow us to reject slave morality and avoid *ressentiment*? We have seen that resentment is an attitude that develops following the conviction that the future is circumscribed (e.g., the masters will always be dominant) and is bound to be contiguous with the past and the present—hence the title of this book, *Chronopathologies*. Deleuze is pointing toward the way in which this final synthesis of time is about pure difference, whereas the other two aspects of the temporal have a less fundamental conception of difference. The eternal return affirms the excessive and the unequal, the multiple and the different. It is about change, or "difference as the origin, which then *relates* different to different in order to make it return as such" (DR 125), and for him this reversal of the tradition has ethico-political consequences as much as ontological ones. We need to seek encounters in which the anticipatory (and potentially judgmental and moralizing) aspects of subjectivity have been stripped away (in fact, as we will see, that even entails encounters without what Deleuze calls the other-structure), so as to allow for the new, and for an event, in the fullest possible sense of the term.

What this might consist in is explored in his collaborative works with Guattari (especially *Anti-Oedipus* and *A Thousand Plateaus*), and Paul Patton and numerous others have more than adequately documented the political significance of these writings.[18] In this book I want to attempt something riskier although not incoherent—a reading (and a problematization) of the politics of time that are at the heart of the philosophies of difference of *Difference and Repetition* and *The Logic of Sense*. While it is clear that one must be careful in leaping to political conclusions on the basis of these texts, it is also clear that Deleuze intends his transcendental descriptions of the time of the eternal return and the future to have a normative and ethico-political import. We can note here, for example, that for Deleuze that which returns in this privileged mode of time is only those beings who have gone beyond the limits of their power, who have transformed themselves and one another. On the basis of this, it can be concluded that he will have reservations about any kind of political philosophy that rests on given identities (especially the self or subject), or that is based on achieving certain ends, perhaps even including attempts to institute a fairer distribution of resources, at least if broader issues are not also taken up. Unlike many Marxists, Deleuze suggests that "the economic is never given properly speaking, but rather designates a differential virtuality to be interpreted, always covered over by its forms of actualisation: a theme or 'problematic' always covered over by its cases of solution." He also adds that "practical struggle never proceeds by way of the negative but by way of difference and its power of affirmation" (DR 186), and suggests that, "revolution is the social power of difference, the paradox of society" (DR 208). Of course, this affirmation of difference is precisely what the time of the eternal return purports to give us, and in this sense it is the transcendental condition of the group politics that he pays more attention to in his collaborative books with Guattari.

The Politics of Derrida and Deleuze

Although there are important differences between these two theorists, we have seen then that they have a shared emphasis upon the future that defies anticipation, and a shared critique of any kind presentism, and there are clear political motivations behind this; when the future is thought to be known, this tends to lead to either Fascism or Communism, in which that future state of affairs can justify the violent means needed to get there. In Deleuze's case, this point is accompanied by a more Nietzschean insistence on the potentially pernicious psychological consequences of subjectivity and its inevitable moralisms (i.e., *ressentiment*). Wary of this kind of thinking, Derrida and Deleuze valorize the aspect of the future that is contentless and which cannot ever be known. Both theorists also affirm time as the most radical form of difference, and insist upon the importance of this difference in opposition to the priority that much of the philosophical tradition has accorded to a temporal "now" moment. Moreover, for Deleuze and Derrida the future is also linked to novelty, to invention, to the emergence of the new, and hence to the singularity of any event worthy of the name—that is, to an event that is not subsumable under considerations of context, historical precedent or structure. However, it needs to be noted that this apparent equivalence between these theorists can be complicated through the recognition that in his eulogy for Deleuze, Derrida hints at a fundamental disagreement with the Deleuzian formulation of philosophy as being concerned with creating concepts. Of course, it is true that Derrida has a more circumspect and qualified attitude to philosophy than Deleuze, considering philosophy to always be partially betrothed to representational thinking.[19] That said, Derrida's discontent with this idea of philosophy as being on about concept creation is not purely about what should be called philosophy (and hence definitional). It is also about the very idea of the creation of the new itself. For Derrida, the new is highly unlikely, improbable, and he continually suggests things like "invention, if there is such a thing" (PF 39). For Deleuze, however, genuine philosophy creates new concepts, and although this may be relatively rare, in his work there is also a transcendental commitment to the new, to difference, that is not accompanied by the obligatory "perhaps" or other forms of literal or implicit scare quotations that Derrida regularly employs. Of course, phrasing the differences between these two French philosophers in such a way runs the risk of making Deleuze look like a naïve utopian thinker. This may seem reductive, but from the perspective of Derrida's work it is also not altogether false. For Derrida, pure difference cannot be experienced because the metaphysics of presence, as well as social habituation, cannot be simply avoided. A time without subjectivity (e.g., *différance*) might produce or make (im)possible subjectivity, but it is, paradoxically, only experiencable in the context of just such a subject. Similarly, the radical futurity of the "to come" is experienced only as a disruption within unity that is necessarily marginal and that is always partly inaccessible. On the other hand, in the work of Deleuze there is a suggestion that any imperialism of the same, or law of self-presence, is far more unstable and, consequently, that the play of difference, even pure difference, can (and ought) be experienced in a modality of futural time that dispenses with both identity and subjectivity.

The distinction between messianisms and the messianic can also help to clarify this important difference between Deleuze's and Derrida's exaltations of the future. For Deleuze, the syntheses of memorial and habitual time are the basic conditions for subjectivity, but subjectivity is radically absent from the futural experience of time: the time of the eternal return of difference. Now, while it is clear that Derrida's notion of the messianic is not reducible to subjectivity but is other than it, there is also a sense in which the messianic relies upon subjectivity and is inconceivable without it. This is because the messianic is best construed as the exposure of a subject to this radical difference and, for Derrida, the subject is also one of those many unities that simultaneously resists the experience of the messianic. Derrida's notion of the messianic hence contains a more psychological register, in that he argues that while we persistently hope for the arrival of the wholly other, the prospect also scares us, and we hence harbor a desire for the coming of the messiah to be indefinitely postponed. Even if it were temporally possible for the messiah to turn up, the subject would resist this, and an experience of pure difference (or pure futurity) hence seems eminently unlikely. Perhaps this should not surprise us unduly given Derrida's enduring insistence upon the impossibility of simply overcoming metaphysics, but it is an important difference between he and Deleuze, and it ensures that subjectivity is never completely thrown into the volcano or dispensed with, as sometimes appears to be the hope in Deleuze's work. The experience of pure difference remains tenable for Deleuze (even if improbable), but this is not so clearly the case for Derrida, who seems to hold true to his earlier formulation, at the start of his long academic career, regarding time necessarily being metaphysical. Derrida accepts and even emphasizes that there is an experience of the future, and of the untimely, but these are only possible in the context of a disruption to a metaphysics of presence, and in which a subjectivity produced by such a metaphysics resists these futural experiences. The ethico-political significance of this difference is only beginning to be examined in detail.[20] Derrida's ongoing vigilance and caution means that deconstruction will never be quite as politically radical as Deleuze and his followers, but that does not preclude it being effective. Deconstruction's ongoing resistance to teleologies of the future is an important guardrail against absolutisms of any and all sorts, although questions remain to be pursued regarding the kind of political calculations—and politics involves compromise—that might best respect this significant *moral* insight at work in deconstruction. In the chapters that remain in part 2, I will look at Deleuze and then Derrida regarding the connections between time and normativity, and seek to proffer what might be called some phenomenological reminders throughout that anticipate the arguments of part 3 against key aspects of these transcendental philosophies of the future.

6

Wounds and Scars: Deleuze on the Time (and the Ethics) of the Event[1]

The wounds of the Spirit heal, and leave no scars behind
(G. W. Hegel)[2]

Let me begin this chapter with a declaration that I will develop and attempt to justify in what follows: Deleuze's oeuvre is best understood as a philosophy of the wound, synonymous with a philosophy of the event,[3] whereas the philosophy of his immediate predecessors in the phenomenological tradition constitutes a philosophy of the scar, with phenomenological and embodied intentionality (including the significance given to habit, coping, and the intentionality of the "I can") resulting in a concomitant refusal to privilege the event as wound. Precisely how Deleuze understands the event and the wound remain to be enumerated, but various consequences hang on this difference, including the divergent ethico-political orientation that undergirds Deleuze's work in comparison to the tacit ethics of *phronesis* (*Verstehen*) that can be ascribed to much of the post-Husserlian phenomenological tradition. Although this wound/scar typology in regard to Deleuze and phenomenology might appear to be premised on a metaphorical conceit, this chapter will show that the motif of the wound recurs frequently and perhaps even symptomatically in Deleuze's texts, in that it often occurs where he is attempting to delineate some of the most important differences (transcendental, temporal, and ethical) between himself and his phenomenological predecessors. While I will ultimately seek to question aspects of his philosophy and ethics of the event, it is worth briefly describing Deleuze's position in a little more detail before the point of my own intervention can become apparent.

Despite the ways in which this wound/scar typology appears to have a primarily spatial register, these terms are explicitly deployed temporally by Deleuze, who consistently describes the aspect of time that his work privileges as caesura, break, cut and wound, something akin to what Elizabeth Grosz' recent book calls the "nick of time."[4] The priority that he accords to this nick of time, this temporal wound, does not follow from a common sense understanding of temporal anteriority, such as the observation that scarification follows from, and is a direct causal consequence of, a prior wound. Rather, Deleuze advocates a renewed understanding of both time and the relation between cause and effect, and, as we will see, the event is explicitly understood as an effect rather than a cause. Without yet explicating this somewhat counterintuitive position that Deleuze develops through an engagement with Stoicism, suffice to say that he intends to problematize any exclusively unidirectional, empiricist understanding of causality, instead pointing to reversals that interrupt the order of brute physical causality, and insisting upon the subsistence of a more subterranean causality that obtains on the level of the virtual/transcendental (the "quasi-cause") and that also haunts and at least partly produces the actual (quasi-causality does not function on the basis of strict causal necessitation and determination). We will soon explore the significance of these orders of causality and their relation to time in Deleuze's *The Logic of Sense*, but to link this expanded understanding of causality to the key argument of this chapter, when Deleuze treats the event as synonymous with the wound, the wound that he is prioritizing is both temporal and transcendental, rather than an empirical event that happens. As we will see, for him, the event never actually happens or is present; it is always that which has already happened, or is going to happen, and it is always already scarified through corporeal and psychological processes of adjustment and introjection. As such, his manner of thinking the relation between wound and scar is not one of empirical antecedent or spatial succession, and unlike the Hegelian epigram with which this chapter began, there is no healing or overcoming of this transcendental temporal wound (the future that is perennially "to come," the pure past that never was) that he continually highlights the importance of.

Without digressing unduly in this regard, it also seems fair to suggest that few twentieth century phenomenologists would endorse Hegel's comment, albeit for very different reasons to Deleuze. Indeed, for the phenomenological tradition it is more accurate to suggest that all healing leaves scars. This includes the scars of historical time, which preclude the teleological understandings of history of both Hegel and Marx, for whom, in different ways, the scars of the dialectic can ultimately be overcome or sublated. Likewise, we should note that the attempted phenomenological reduction is itself necessarily incomplete, at least for those post-Husserlian phenomenologists who, paradoxically enough, provide phenomenological evidence for our inability to access consciousness purely and without remainder. For Heidegger, Merleau-Ponty, and Sartre, among others, the phenomenological "return to the things themselves" is always partially successful and partially aborted (e.g., scarified), in that inner and outer are always coimbricated and contaminated, as some residue of our socio-historical situation—being-in-the-world, and the inevitability of state-of-mind, or mood—is always presupposed. Indeed, it is interesting to note that mood is explicitly described as past oriented in Heidegger's *Being and Time*. No pure self-presence (or therefore a phenomenological reduction) is possible for Heidegger, because of the omnipresence of mood (or attunement as Joan Stambaugh translates it), as well as the priority that Heidegger argues that *Dasein* must accord to the "not yet," the future.

Of course, Heidegger's work hence presages important aspects of the poststructuralist account of time (including Deleuze's), but the importance that existential phenomenology accords to embodied comportment (and the significance of coping and the ready-to-hand for Heidegger, equilibria for Merleau-Ponty) is also, arguably, accompanied by a priority accorded to that which makes the world meaningful and a temporal binding or gathering as we will see in more detail in chapter 9. More specifically, they are concerned with the way in which time scarifies, conjoins rather than disjoins, as well as by the way in which lived time almost inevitably forms a neat and unified continuum. In *Difference and Repetition*, Deleuze makes this explicit in his critique of the temporality betrothed to good and common sense (the *ur-doxa*), which he argues continue to afflict phenomenology.[5] Rather than dwell upon this critique here, however, this chapter will be devoted to a sustained consideration of his own more positive account of time and the wound in *The Logic of Sense* and, to a lesser extent, in *Difference and Repetition*. This will involve explicating his understanding of the event, as well as the notoriously opaque ethics of counter-actualization that are bound up with it, before raising certain problems that are associated with the transcendental and ethical priority that he accords to the event and what he calls the time of *Aion*. I will conclude by proposing a dialectic between the two aspects of time that he counterposes (*Aion* and *Chronos*, roughly the disjunctive and the conjunctive) that does not instantiate any kind of a priori privilege of the one over the other.

Aion: The Event as Wound of Time

The theme of the wound is arguably one of the key aspects of Deleuze's enigmatic book, *The Logic of Sense*, but in order to understand its centrality it is first necessary to enumerate a series of interrelated distinctions that undergird this text. Deleuze's strategy in this text and others is to begin with something that seems to resemble a typical opposition, but which is soon shown to be an interrelated one, in that the terms involved cannot be considered complete without each other and some kind of process of becoming holds between them. The important concepts for our purposes are: *Aion* and *Chronos*, surface and depth, wound and scar, event and state of affairs (and, imported from elsewhere, the virtual and the actual). The displacements and overlappings that occur between all of these terms is one cause of the difficulty of the text in question, but although they are certainly not synonyms (the *Aion/ Chronos* distinction pertains to time whereas surface/depth refers primarily to space—both are involved in the paradoxical constitution of sense), I will suggest that these terms nonetheless all have an isomorphism of function that maps on to the overarching distinction between the virtual and the actual that preoccupies Deleuze throughout his career. To schematically define these latter terms, the virtual refers to that which is creative, productive, and transformative (a transcendental field of difference), whereas the actual refers to that which is created, produced, and of the realm of identities, sameness, and all that currently *is*.[6] In what follows, I will argue that these distinctions also have an order of priority embedded within them, in that Deleuze consistently associates *Aion*, surface and wound (all of which are of the order of the virtual) with the event, or at least with the "truth of the event." In the section of the chapter following that, I will also show that *Aion*, surface and wound are directly tied to the possibility of a new ethics, even if it is not an ethics that Deleuze himself thoroughly explicates.

By way of beginning then, it is not particularly controversial to suggest that in *The Logic of Sense* there is a privilege given to surfaces over depths, as well as to that which breaks open the surface but is nonetheless not opposed to it—the "crack," the wound, which he consistently insists does not reveal something deeper and more fundamental. Although it is a difficult book in which to discern traditional philosophical arguments, he justifies this temporally, transcendentally, and in relation to the event that does not occur in regular linear time. As such, the priority that Deleuze gives to the surface is intimately bound up with the temporal priority that he gives to the time of the event, and the pivotal distinction that he draws between the time of *Aion* and the time of *Chronos*, terms that Deleuze also deploys with Guattari in *A Thousand Plateaus*. In the more avowedly Stoicist understanding of *The Logic of Sense*, *Aion* is described as the time that constantly decomposes into elongated pasts and futures, whereas *Chronos* is said to be composed of a series of interlocking presents. *Chronos* measures temporal actualization and the realization of an event—its "incarnation into the depths of acting bodies and its incorporation in a state of affairs" (LS 73). Deleuze explicitly suggests that this latter time (which involves incorporation, mixtures, and depth; all figures of mediation) is the time of the scar (LS 10), the realm of the actual, including bodies and states of affairs. It is, however, the incorporeal surface, rather than the corporeal depth, that Deleuze associates with the privileged time of *Aion*, which subdivides endlessly into the past and future, and the event that likewise never actually occurs in present time. The time is never present that allows for an event to be realized, or to definitively exist. On this understanding, the event is "always and at the same time something which has just happened and something which is about to happen; never something which is happening," never an actuality. It subsists rather than exists. Whereas Deleuze suggests that *Chronos* "is cyclical, measures the movement of bodies and depends on that matter which limits and fills it out" (LS 64) ,*Aion* is a "pure straight line at the surface, incorporeal, unlimited, an empty form of time, *independent of all matter*" (LS 73). It is important to note, therefore, that the time of *Aion* is independent of both matter and the present, which on Deleuze's understanding means that it is independent of habit, and thus of embodied forgetting, coping, and the maintainance of equilibria that is typical of bodies and states of affairs (it is also independent of reason, as we will see, although that need not suggest that it is independent of the concept or the virtual Idea). This is a curious and paradoxical thought, especially given the materialism that Antonio Negri associates with his (and Michael Hardt's) reworking of this Deleuzian understanding of time,[7] and it is one that clearly has a distinctive understanding of the event betrothed to it. We will come back to the question of this temporal independence from matter, a line of inquiry that is shared in different ways by both Deleuze and Derrida,[8] but suffice to say that for Deleuze there is a gap between the transcendental and the empirical that refuses all forms of mediation. That gap or wound is time, and in positing this temporal independence that is at the heart of what he describes as his "secret dualism" (LS 4), there is a sense in which Deleuze's work appears to be premodern, or pre-Kantian, despite the fact that there are also various ways in which his work can be understood as a radicalization of the Kantian conception of time and the transcendental.[9] The significance of time as a transcendental condition will concern us throughout, but to put Deleuze's point more precisely, the wound that refuses mediation and recuperation (scarification) involves the nonpresentist and more paradoxical manifestations of time (we can think here of the discussions of *Aion*, of "time out of joint," the pure past that never was, the eternal return of difference as empty form of time, etc.).

We have seen that there are various conceptual oppositions drawn in *The Logic of Sense*. It is important to reaffirm, however, that there is also an asymmetrical reciprocal determination that obtains between these myriad oppositions, in that they are modeled on the virtual/actual distinction, and in the latter regard Deleuze insists that the actual ramifies back upon the virtual, albeit in a manner that he gives less attention than the reverse determination from the virtual to the actual.[10] Certainly Deleuze insists upon the necessity of relations between the actual and the virtual, and, at times, a similar logic pertains to the distinctions between *Aion* and *Chronos*, surface and depth, and event and state of affairs. Nonetheless, and despite the odd Deleuzian disclaimer, it is difficult to dispute that these distinctions also involve a hierarchy and order of subordination, with the transcendental or virtual term of the pair prioritized in both a philosophical and ethical sense. In regard to the distinction between surface and depth (and thus between the incorporeal event and the actions and passions of bodies), we can note that in relation to depth Deleuze comments that it is through "infinite identity that contraries communicate" (LS 200). Written just a year previously, we know that the entirety of *Difference and Repetition* is devoted to undermining the primacy of such a philosophy. In *The Logic of Sense* itself, we are told that the realm of the surface, and the counter-actualization associated with it, are already "infinite distance" (LS 201). Similarly, Deleuze's descriptions of the "mono-centered" return of *Chronos* betray his attitude toward this aspect of time, and we must note that he also links *Chronos* directly to both nouns and equivocity (LS 211). The latter is not genuine ontology, for him, as it does not attain to the order of univocity (and verbs in the terms of *The Logic of Sense*; the eternal return of difference rather than sameness in *Difference and Repetition*), which he consistently proclaims as the only coherent ontology. If *Chronos* and depths are comparatively disparaged, *Aion* and the surface are, on the other hand, described as "the transcendental field itself, and the locus of sense and expression" (LS 142). As we will shortly see, the time of *Chronos* is also closely (if not inextricably) associated with an ethics of *ressentiment*, whereas *Aion* and the surface are fundamental to any affirmative ethics that accepts Deleuze's basic Stoic-inspired premise: "either ethics makes no sense at all, or this is what it means and has nothing else to say: not to be unworthy of what happens to us" (LS 169).

Before we get to the question of ethics, however, it is necessary to comprehend Deleuze's repeated suggestion that events are only effects, and this will require consideration of his paradoxical understanding of cause and effect (LS 10, 29, 241), in particular the manner in which what he calls the "quasi-cause" comes to stand in for something closely related to virtual-actual causation. How then, can an event be said to be an effect? According to Deleuze, the reversals experienced by Lewis Carroll's famous character, Alice, are said to give us some kind of indication. Notably, Alice is punished before having done anything empirically wrong, and she cries before pricking herself. Now, there may be empirical explanations for such behavior, but these are not of the order of the event. To focus upon them would leave what he calls the expressive aspect of the event untouched, which does not obey the logic of our anticipations and rational reconstructions and predictions. Empirical explanations follow from a reality that is both *Chronos* and *Aion*. Moreover, they typically fail to see that what has occurred is never wounding because of any particular actuality, whatever it may be, but that we are wounded because of the prospect of the future being the harbinger of further unknown monstrosities, or because of the relation that any given actuality bears to the complex of temporal syntheses that is our past, noting (again) that this memorial past synthesizes from passing moments a form that never existed before that operation. To phrase this more positively, we are also likewise inspired and touched by events because of their relation to the time of *Aion*. As such, it is the future and the past that touch and wound us; they are the time of the event. According to Deleuze's understanding of this period (prior to the greater influence of Guattari), psychoanalytic explanations were capable of understanding this and thus promised to provide the science of the event. And it is easy to see why he might have been impressed with psychoanalysis. Despite Freud's scientific pretensions, psychoanalysis need not require that there actually occurred a wounding primal scene that the child and adult subsequently adjust to (this is partly why Karl Popper famously calls it unfalsifiable, and, without buying into a Popperian objectivism, we will consider the philosophical efficacy of the particular techniques of transcendental argumentation that Deleuze deploys). While Deleuze calls into question the normalizing tendency that he finds at the heart of psychoanalysis (hence his positive discussion of the "perverse" structure toward the end of *The Logic of Sense* which we consider in the next chapter), for him psychoanalysis nonetheless managed to grasp the event as an effect, and one that does not simply follow from any single cause, or from a concatenation of actual causes that constitute any given state of affairs.

This is because, for Deleuze, the event is subject to a double causality (a double structure), one aspect of which involves a mixture of bodies and states of affairs (e.g., empirical causes), the other aspect being the conflagration of different incorporeal events that are of the order of the quasi-cause (LS 108). Psychoanalysis, too, has a double structure or a secret dualism at play in it, in the manner in which the symbolic order of language is disjunct from the realm of bodies. Deleuze endorses aspects of this account in the latter stages of the book, and his key point is that the event also results from the quasi-causal interaction of other events in the realm of the virtual, which is associated with *both* past and future contra Alain Badiou's interpretation that associates the virtual rather exclusively with the past.[11] The virtual relation between different incorporeal events is not a relation of causal necessity; on the contrary, it is a relation of expression (LS 194). Event-effects are said to assume among themselves a relation of expression that is quasi-causal. As such, the event haunts and subsists without inhabiting bodies or places in a manner that is closely related to the unreal and ghostly causes that feature in Derrida's *Spectres of Marx*. Exactly how this relation differs from that of ordinary causation is not spelled out in any detail because Deleuze suggests that this quasi-causal relation with the virtual/transcendental is unknowable. We can know that it occurs, but we cannot trace particular chains of quasi-causation.

While Deleuze acknowledges that the event does have immediate corporeal causes (LS 10), and states of affairs that precipitate the event, they
are not sufficient causes of the event itself. Explicating the event at the level
of corporeal cause(s) and historical conditions always leaves something untouched. For example, there may be a momentous historical event (let's say
May 68) brought about by certain preconditions that can be fairly rigorously
delineated, but that actual state of affairs is not, for Deleuze, commensurate
with the Event of May 68. As an effect, the event is always that which has just
happened or is about to happen, but never of the order of that which currently
is happening. He hence associates the event with reversals between the orders
of past and future, but not the present.[12] The event is thus virtual, perhaps
even extra-worldly. It can never come about, but produces and conditions
that which comes about. According to Peter Hallward's provocative interpretation, this means that Deleuze's work is not sufficiently concerned with the
world and history.[13] While such an interpretation arguably ignores the double
causality that is at work in the architectonic of Deleuze's philosophy, it nonetheless seems to me that Hallward is right in one important sense: the emphasis that Deleuze gives to *Aion* over *Chronos* means that he also prioritizes
the effect as quasi-cause (the virtual) over the cause (actual) itself. There is a
sidelining of ordinary causality (*Chronos*, bodies, states of affairs, mixtures,
depths, etc.) as Hallward complains. To put it another way, the normative
emphasis upon creativity and difference that undergirds Deleuze's work tacitly devalues "average everydayness" (and thus scarification and mediation),
a term that is taken from Heidegger's *Being and Time* and the privilege this
book initially accords to the ready-to-hand and a pragmatic coping with our
environment that is the basis upon which the abstractions of the present-at-
hand (reflective analysis) are based, but that can also be discerned having a
normative impetus in the work of the early Merleau-Ponty (think of the "I
can" of embodied coping that plays such a major role in *Phenomenology of
Perception*). Although couched in a widely divergent vernacular that does not
so clearly reinstitute a form of metaphysics, a similar move to Deleuze's rejection of this is also made in Derrida's philosophy, one that is again justified
by recourse to a transcendental (or quasi-transcendental) conception of time,
as we will see in chapter 8 and throughout part 3 of this book. One important
question for consideration here hence concerns the validity of transcendental
argumentation in the specific manner they are deployed by Deleuze. Indeed,
in a different context, James Williams asks of this Deleuzian conception of
the transcendental: "how far can Deleuze's insistence on the reciprocal determination of condition and conditioned be maintained in the face of the
objection that this quasi-causality is a fictional distortion of scientific causal
regularity?" Shortly after, Williams goes on to suggest that the important

question becomes: "what arguments does Deleuze have for the validity of transcendental deductions that are not based on pure beginnings?"[14] In chapter 10 I coin the term "empirico-romanticism" to express reservations of this kind about Deleuze's position, which involves the simultaneous moral and transcendental positing of some kind of radically disruptive event, akin to a first genetic cause, that forces us to acquire new faculties and new modes of being. The more general issue becomes one regarding the limits and strengths of transcendental philosophy in its various guises, and in Deleuze's case there seems to be a genuine risk that the centrality accorded to his particular conception of the transcendental tacitly devalues ordinary causation of both a scientific and embodied bearing; it also results in a immanently unjustified hierarchy and an associated ethics.

Let us think, however, about what this means for the wound. Wounds are also incorporeal for Deleuze and hence not part of ordinary causation: instead, they are regularly treated as synonymous with the event. For example, he comments: "the Event itself, the result, the wound as eternal truth" (LS 51); "why is every event a kind of plague, wound, war or death?" (LS 172); "is it the case that every event is of this type—forest, battle and wound—all the more profound since it occurs at the surface?" (LS 12). Some of these comments are bound up with his discussion of Joë Bousquet, to whom he attributes the Stoic maxim: "my wound existed before me, I was born to embody it" (LS 169). He also makes various related observations in his engagement with Charles Péguy's theory of the event in *Difference and Repetition*, but the frequency of allusions to the connection between the event and the wound suggests that they are also intimately connected for Deleuze himself. Now, we know that the wound cannot be understood as something that accidentally and contingently befalls us. That would be to treat it as an empirical event, rather than of the order of the virtual, the event-effect. But what might a virtual-wound be? If we look to pick out particular things as exemplars, we commit a transcendental mistake: we confuse something like Being, or perhaps more aptly, becoming, for actual beings. Granted that this is so, one cannot help but wonder what is surreptitiously being imported into the equation by the naming of the virtual event-effect as wound and opposing that to the scarified realm of bodies and recuperation. My questions for Deleuze then, are threefold. First, is this event-wound priority tenable? Second, can an ethics be deduced from this transcendental priority? And third, what ensues (consequentially) when a transcendental and ethical priority is given to the non-embodied, to the virtual wound that is independent of all matter? Does it mean that his avowedly political work runs the risk of degenerating into an eternally patient moral perfectionism, which eschews both rational calculation (as well as the basic causality of bodies) in favor of stylized prophesies and transcendental dreams of the disruptions of the past and future? Although

it is a difficult question as to what one means and expects by the term "the political," I think that this risk is present in Deleuze's work when the transcendental aspect of his various oppositions is so frequently prioritized in a quasi-
causal sense and the manner in which the actual reciprocally impacts upon
the virtual is given comparatively scant emphasis and attention. Although the
event and states of affairs may be in a causal relation for him, it is another
causal relation (the interaction between different event-effects in the virtual)
that is given the greater importance by Deleuze, despite being unknowable.
This becomes clearer upon consideration of the opaque and difficult program
for a new ethics that he proposes.

Counter-Actualization: An Ethic of the Wound

The full picture of Deleuze's ethics is rather more difficult to grasp than is
usually assumed. Frequently, we are told that his work involves an immanent
ethics that synthesizes the respective ethics embedded in the Nietzschean
and Spinozian philosophies that eschew transcendent judgment as well as
rule-based accounts of ethics (or law).[15] While this is at least partly true,
such a characterization does not adequately come to grips with an aspect of
Deleuze's work that is most apparent in *The Logic of Sense* (but is also in
evidence elsewhere), and that pertains to the way in which this immanent
ethics cannot really be understood as immanent to the actual physico-biolog-
ical world, but is more aptly said to be contrary to it, or at least contrary to
the dominant aspect of it (here is where the Bergsonian distinction between
the virtual and the actual becomes particularly important to his position).[16]
Understanding this relies upon at least some comprehension of the obscure
ethos regarding the time of *Aion* that is central to the text. Indeed, it is arguably only upon seeing the manner in which counter-actualization (and deter-
ritorialization) is bound up with the time of *Aion* and the virtual, along with
comprehending the priority that Deleuze accords to each because they are
"independent of all matter," that we can begin to grasp what an ethics of the
event-effect might be.

Before attempting to answer this question, however, it is worth first
guarding against the objection that Deleuze does not draw any ethical conclusions on the basis of the transcendental priority that he accords to *Aion*
and to the surface (wound). Suffice to say, such an interpretation is very difficult to maintain in the face of the *The Logic of Sense* itself, even without
considering series 21 and 22, which explicitly consider the question of ethics. While Deleuze does not offer any moral prescriptions, there is a clear
normative force at work in the distinction between *Aion* and *Chronos*, hence
persistent rhetorical questions such as: "is there not in the *Aion* a labyrinth
very different from that of *Chronos*—a labyrinth more terrible still, which

commands another eternal return and another ethic (an ethics of Effects)?" (LS 72). Despite the fact that this appears to be an open question, his invocation of "*another* eternal return" here is also important. We know that there is both an ontological and normative force given to the time of the eternal return of *difference* in *Difference and Repetition*. The eternal return of the *same*, on the other hand, is described as being of only the most simplistic and introductory value. Without considering the warrant for this as an interpretation of Nietzsche, something very similar is also going on in *The Logic of Sense*. In this book, Deleuze not only calls for "another eternal return," but explicitly associates the wisdom of the actual cause with the eternal return of the same, and a moral eternal wisdom that he denigrates. There are innumerable other comments from Deleuze which suggest that it is an ethics of effects that he is ultimately interested in and the more committed to. He comments, for example, that "Paul Valery had a profound idea: what is most deep is the skin. This is a Stoic discovery, which presupposes a great deal of wisdom and entails an entire ethic" (LS 12). We will come back to this purported entailment, but he likewise advocates the profound link between the logic of sense and an ethics (LS 38). While Deleuze proffers no prescriptive or rule-based account of ethics, and while this invocation of another ethic (an ethic of *Aion*) does not suggest that we can simply dispense with an ethics of *Chronos* and the depths (rules, rationality, causal considerations), there is in *The Logic of Sense* (and in Deleuze's work more generally) a priority given to this ethics of *Aion* (an ethics of nonpresentist time), which is synonymous with an ethics of wounds. I will try to develop these claims via a sustained discussion of Deleuze's appropriation of the Stoic ethic of willing the event, and the inflection that his concept of counter-actualization gives to this idea.

Deleuze suggests that Stoic ethics oscillates between two poles: on the one hand, between advocating the greatest possible participation in a divine vision that gathers in depth all of the physical causes; on the other hand, it is also concerned with willing the event (surface), whatever it may be, and without any interpretation or intent to integrate it within the unity of all physical causes. Deleuze suggests that the first Stoic pole is problematic because of the physicalism that it presupposes. For him, events differ in nature from the corporeal causes from which they are the result; they have other laws (clearly not deductive-nomological ones) and other incorporeal forms of relation (i.e., quasi-causal, expressive) (LS 163-4). Although it is arguable that he simply replaces the Stoic impetus toward a gathering in depth of all physical causes with an affirmation of the univocity that obtains between virtual event-effects, it must be noted that on his account, "the unity of events or effects amongst themselves is very different from the unity of corporeal causes amongst themselves" (LS 75). It is, however, the second pole of the Stoic ethic with which he is primarily concerned and which he wishes to rework for his own purposes. Accomplishing this second aspect—willing the surface event without interpretation—depends pivotally upon one's relation to time, because it is not a matter of simply willing all that befalls us. That interpretation of the *amor fati* is insufficient and amounts to a form of indifferent resignation that Deleuze finds to be ethically problematic. As John Sellars puts it, it is a human Stoicism that tacitly remains resentful, rather than a cosmic Stoicism that involves both affirmation and a more paradoxical relation to time.[17] Indeed, according to Deleuze, the genuine Stoic sage must simultaneously wait for the event as something eternally yet to come and always already passed (*Aion*) (LS 166). The truth of the pure event is this paradoxical temporal dimension of it: the waiting without expectation of an arrival that is quite closely related to what Derrida calls the messianic, along with the recognition that the event has also already passed. While Deleuze also argues that the sage wills the actualization of the event-effect and the giving body to the incorporeal effect, even then the sage ought to will not exactly what occurs, whatever it may be, but something in that which occurs (LS 170). We have already noted that Bousquet serves as an example of what this might involve. As Sellars comments, "the task for Bousquet was to transform the event of the wound from a tragic external assault that afflicted him into a vital and necessary event in his life that made it possible for him to discover himself as a writer, to become who he already was."[18] We can conclude from this that despising any particular wounding-event is a form of *ressentiment*, as, to a lesser extent, is the traditional Stoic ethic of expecting suffering and misfortune but soldiering on (both make a transcendental mistake when they treat the wound empirically). On the other hand, embracing the event and the transformations it induces (not the brute actuality) is *amor fati*. This is the rather stark alternative that Deleuze seems to leave us with, and he goes on to encourage us to become the offspring of one's events, not of one's actions. To become the offspring of the virtual intensities that subsist in oneself, which is another way of saying to express the wound and to make it the quasi-cause of one's life. Now this cannot mean to become the offspring of one's emotions or passions, or even to intensify one's emotions and passions. After all, emotions are, for Deleuze, bound to a subject, and passions are considered to be of the order of a state of affairs rather than of the order of the event. (LS 7) It is difficult to pinpoint positively what his ethics might involve, but it seems to require the recognition that we are all traversed by some kind of fault-line (a virtual, impersonal intensity) that is supra-individual and not confined to the realms of bodies and states of affairs. Whether this wound can be distinctive for each of us or ultimately partakes in one transcendental wound is not clear, but it is the concept of counter-actualization that he uses to more fully describe what is involved in the appropriate manner of giving body to an incorporeal event-effect.

In describing his ethic of counter-actualization, Deleuze suggests that each time the event is inscribed in the flesh, "we must double this painful actualization by a counter-actualization which limits, moves and transfigures it" (LS 182). Inscribing the event in the flesh (in the realm of bodies and habits) is hence necessary for the sage and for ethics, but it is not the key aspect of his ethic (LS 192). Rather, it is the potentialities of that actuality that are expressed, not merely the literal re-inscription of the same;[19] not pathological repetition, but repetition with a difference. Counter-actualization must hence "limit, move, transfigure" and mime that which effectively occurs. While the event is brought about by the living-present, by bodies, states of affairs and reason, its eternal truth is irreducible to them. The event is the result of the actions and passion of bodies, but his ethics affirms its irreducibility to these origins, done by linking it to a transcendental quasi-cause (wound, *Aion*) rather than the empirical cause. This is, as Deleuze himself acknowledges, an ethics of the mime and of acting. Sensations and intensities can be extended beyond the singular through the expressive and dramatic practices of counter-actualization. For him, it is "the free man, who grasps the event, and does not allow it to be actualized as such without enacting, the actor, its counter-actualization" (LS 173). This counter-actualization involves a delicate operation, in that we need to limit ourselves to the counter-actualization of an event (and thus embrace our wounding virtual-effect) without the full actualization of this wound that characterizes the victim and the patient. For Deleuze and Bousquet alike, the wound exists before us, before any particular subject or individuality, and yet we are born to embody it, thus "becoming the quasi-cause of what is produced within us" (LS 169). Again, it is difficult to understand precisely what this means, but we know what it does not mean. We should not be indifferently resigned to whatever happens to us (as in the commonly received understanding of Stoicism), and it is also important to note Deleuze's second and inverse warning: "counter-actualization is nothing, it belongs to a buffoon when it operates alone and pretends to have the value of *what could have happened*. But to be the mime of *what effectively occurs*. . . is to give the truth of the event the only chance of not being confused with its inevitable actualization" (LS 182). There are then, two main ways of misunderstanding and mistakenly living his ethics of counter-actualization: assenting to whatever actually happens indifferently and with resignation; flippantly miming other possibilities that bear no effective relation to what happens. No prescription can tell us how to accomplish this, but we can see that counter-actualization endeavors to achieve that most paradoxical of things: to express and even illicitly embody the virtual, to feel that time which is not. If the present (*Chronos*) measures the temporal realization of an event, and the way in which the wound is covered over and incorporated into a state of affairs, counter-actualization depends upon maintaining a relation to time that opens itself to the immemorial past (that past that defies conscious memorial reconstruction) and the future that is to come, a time that retreats and advances, divides endlessly into a proximate past and an imminent future. This is the time of *Aion*, the wound of time.

How can an ethics be based on time, and on the aspect of an event that never actually occurs but is understood as something in that which occurs, and which is also said to be both always already passed and yet to come? Well, we have already seen that Deleuze is following in the footsteps of Nietzsche. What Nietzsche diagnoses as *ressentiment* (a disgust for life that trades in negativity) is a taking revenge against the fact that time passes. From the perspective of some particular present, we might rail against suffering and injustices, whether they be anticipated or endured. The problem with this attitude, however, is that it treats the wound-event as somehow wholly outside of us. Suffice to say that this is the reverse of what Deleuzian counter-actualization aims to achieve. It wants the wound to give birth to us, but not to be the same as us. Deleuze insists that there is no other ill-will than *ressentiment* of the event, and, given that we also know that for him the truth of the event is *Aion*, we can conclude that there is no other ill-will than *ressentiment* of *Aion*. For counter-actualization to be successful, although it cannot simply return to the virtual it must both show the manner in which the virtual and the time of *Aion* breathe life into that which occurs, as well as simultaneously allow this to happen (this is the tension between the "is" and the "ought," the descriptive and the normative that Deleuze's philosophy negotiates and sometimes conflates). What does that mean for the role of the present, for *Chronos*, for bodies, states of affairs (including empirical wounds and suffering), and even for reason, which Deleuze also suggests is a being of the present? They are all insufficient for an ethics, and his point is not merely that some kind of dialectical relation with an ethics of *Aion* needs to be recognized in order to balance or moderate the monopoly that an ethics of *Chronos* has hitherto enjoyed. Rather, his point is once again both transcendental and normative: the time of *Aion* and the virtual are the condition for the event, and from them he also derives the normative principle of counter-actualization.

The important question then becomes the following: can a transcendental condition also entail a normative principle, even one as opaque as this ethic of counter-actualization? For Deleuze, the transcendental needs to provide the conditions for real experience. If we grant for a moment that his philosophy accomplishes this in its descriptions of the molecular, difference-in-itself, the virtual, the Aionic component of time, etc., in what sense does an ethics of counter-actualization follow from this? It is not clear that it does. Nor is it clear why *Aion* and the truth of the event need to be understood as "independent of all matter." One would have thought that the transcendental condition, the realm of the virtual, is never wholly independent of matter;

indeed, by Deleuze's own lights as evinced by the concept of reciprocal determination in *Difference and Repetition*, it is not.[20] Does this independence of matter, this "secret dualism" wherein *Aion* and the event are privileged, simply reproduce itself at a moral level, with a moral hierarchy? It seems to me that it does. If so, this is philosophically problematic in itself, but there are also reasons to question this ethic in its own right. Because it parallels the movement of the quasi-cause and is associated with the virtual, an ethics of counter-actualization necessarily resists the imposition of any form of transcendent criteria. As such, using this ethics to discriminate between different modes of existence is exceedingly difficult. While this problem is partly overcome in other of Deleuze's texts, especially those on Spinoza and Nietzsche, it is important to note that this ethic also (illegitimately in my view) consigns coping, equilibria, and the body of depths to a secondary status. So what, one may ask? Well, this is, I think, ethically problematic. As has been observed before, albeit for more avowedly political reasons, it is aristocratic,[21] which is also, perhaps, another way of saying, "extra-worldly." This is a strange and counter-intuitive consequence for a philosophy of immanence, and it seems to me that it arises from competing tendencies in Deleuze's work that are never satisfactorily resolved: that is, his post-Kantian philosophy of time and the transcendental (which intercedes intermittently in his ethics), and his immanent Spinozan ethics of immanence (which is avowedly also ontological), the latter of which should theoretically do away with the hierarchies that his transcendental philosophy of time tacitly depends upon.

Difference and Repetition and *What Is Philosophy?*

Although these aspects of Deleuze's work (i.e., the transcendental priority given to *Aion* and the virtual, and a resultant ethics) are foregrounded in a distinctive and perhaps especially problematic way in *The Logic of Sense*, some similar positions continue to prevail in the very different idioms of *Difference and Repetition* and *What Is Philosophy?* I have already suggested an interpretation about the relation that *The Logic of Sense* bears to *Difference and Repetition*. In *The Logic of Sense*, the incorporeal wound is the wound of time, but not of all time understood as some kind of whole; rather, it is the wound of a particular disjunctive aspect of time—*Aion* rather than *Chronos*. More particularly, we have seen that *Aion* is composed of a simultaneous movement in two directions, opening upon both the future and the past. In the terms of *Difference and Repetition*, *Aion* hence encompasses two different temporal syntheses, memorial and futural. In this text, as we saw in the previous chapter, it is only the latter of which that Deleuze understands as inducing a caesura that fractures the "I," and which he hence suggests, "must be determined in the image of a unique and tremendous event." This futural time, exemplified for him by the idea of the eternal return of difference, simultaneously conditions and undermines both habitual and memorial time, and cannot itself be reduced to its corporeal conditions.

In order to understand the priority that he gives to the future synthesis of time in *Difference and Repetition*, it is worth reflecting on what Deleuze has to say about Péguy and Kierkegaard. Both of them are described as great philosophers of repetition, yet he suggests that they were not ready to pay the necessary price, that is, embracing this radical futural wound. For him, "they entrusted this supreme repetition, repetition as a category of the future, to faith . . . However, faith invites us to rediscover once and for all God and the self in a common resurrection . . . they realize Kantianism by entrusting to faith the task of overcoming the speculative death of God and healing the wound in the self" (DR 95). In other words, his objection to their philosophies of repetition is that they heal this temporal wound, cover it over, in a very different manner to their predecessor Hegel but with nonetheless the same result. With Nietzsche's death of god and the wars of the twentieth century, however, things have changed on the philosophical scene. With post-Husserlian phenomenology the healing is never complete and perfect but always scarificatory.[22] As such, phenomenology can be plausibly characterized as tacitly presupposing an ethic of scars (coping), a phronesis that mediates, or a wisdom that searches for the middle.[23] In a sense, it is hence hard to disagree with Deleuze's suggestion in *Difference and Repetition* that it constitutes a philosophy of common sense, of the urdoxa, but the important question is whether common sense is automatically worthy of condemnation and warranting replacement with becoming a "little alcoholic, a little crazy. . . just enough to extend the crack" and in his conviction that "health alone does not suffice" (LS 179). He comments:

> If one asks why health does not suffice, why the crack is desirable it is perhaps because it is only by means of the crack and its edges that thought occurs, that anything that is good and great in humanity enters and exits through it, in people ready to destroy themselves—better death than the health which we are given (LS 182).

While Deleuze may be right in his premise that health alone does not suffice, his conclusion—"better death than the health which we are given"—does not, it seems to me at least, follow. His point in making such a comment is presumably to denigrate a degraded sense of health in which we become trapped and limited, but once more his ethico-political injunction derives much of its force from the presentation of a forced dilemma and an opposition (not a paradox)—either limitation or its overcoming—without due consideration of myriad positions in between. While it is certainly true that we do not want what Deleuze refers to under the label of the perverse-structure (i.e., the pure surface, but also depths being broken open in the manner of Artaud, masochism, death-instinct, etc.) to be covered over by the supposition of rationally self-interested agents, there is also something about the reverse move that is sometimes evinced in Deleuze's work that is unsatisfying and fraught with philosophical difficulties.

Clearly I remain unconvinced by Deleuze's answers, at least insofar as there is a privilege granted to the crack (wound), madness, and even the virtual event-effect, but based on the following we might risk the following epochal formulation: time (*Geist*) and history heal all wounds for Hegel; God heals the temporal wound for Peguy and Kierkegaard; for the post-Husserlian phenomenologists time imperfectly scars; with post-phenomenology (in particular Deleuze and Derrida) it is the wound of time itself that is revalued in a transcendental move that tacitly diminishes the scar. Against a certain postmodern reception of Deleuze as a philosopher of the body, I might even suggest that this transcendental move also diminishes the actions and passions of bodies. Both his transcendental philosophy of time and his associated ethics disparage the imperfect corporeal healing that is always, already at work, just as we will see in chapter 11 that Derrida also endorses and deepens Jean-Luc Nancy's reservations about "ideologies of the body," by stating in *On Touching* that Nancy's worries apply to all philosophies of the body per se. Although this kind of critique is sometimes well-placed, arguably both Derrida and Deleuze's philosophy throw the baby out with the bath water. In other words, the fact that philosophies of the body have sometimes, or even often, been problematic in this way, does not show that their considerations are either philosophically unimportant, or should instead be consigned to the lesser term of a dualism as I argue surreptitiously happens in both of their major works.

Indeed, this tendency is still apparent in *What Is Philosophy?* where Deleuze and Guattari offer consistent refrains against any recourse to the "lived body," their targets here clearly being the existential phenomenology that preceded them on the French scene. They also recapitulate many of these ideas and the motif of the wound and the scar return. Deleuze and Guattari suggest that it is the conceptual personae that counter-effectuates the events, who wills war against past and future wars, who wills "the wound against all scars." Moreover, in redescribing the famous virtual and actual distinction, they also suggest: "from everything that a subject may live, from its own body, from other bodies and objects distinct from it, and from the state of affairs or physico-mathematical field that determines them, the event releases a vapor that does not resemble them and that takes the battlefield, the battle, and the wound as components or variations of a pure event."[24] As such, Deleuze and Guattari reaffirm that the event is actualized or effectuated when inserted into a state of affairs, but counter-actualized or counter-effectuated when abstracted from states of affairs so as to isolate its concept. There is hence still a sense in which one needs to disembody from states of affairs, extract themselves from a lived situation in order to embody the incorporeal event and to experience the counter-forces that might have been, and, in a certain paradoxical sense, nonetheless still are. While the point is arguably not to take oneself "out of this world" as in the title of Hallward's recent book, but rather to live the event in the world (noting that "world" must be understood in an expanded and non-empirical sense) in a way other than it first presents itself, the spirit of the injunction is nonetheless to be true to the aspect of the wound (event) that does not and cannot appear in the world (it is time out of joint, that which ruptures any worldly or bodily sense of integrity and/or unity).

Now it might be protested here that Deleuze's indebtedness to empiricism and his sustained discussions of habit complicate this claim that he marginalizes the actual, bodies, scars, etc., and in a certain sense it does. For him, habit is fundamental to the constitution of subjectivity, as is clear as early as *Empiricism and Subjectivity*, and also in *Difference and Repetition*. His analyses are highly acute in this regard, but it is important to recognize that habit is nonetheless the lesser (ontologically) of the three syntheses of time that he describes in chapter 2 of *Difference and Repetition*. While in a strictly logical sense there cannot, for Deleuze, be a subject without habit, it is the motif of binding that dominates his descriptions of this synthesis of time, and we must note that in the context of his discussion of the three syntheses of time, Deleuze says of the habitual synthesis that "a scar is the sign not of a past wound but of 'the present fact of having been wounded': we can say that it is the contemplation of the wound, that it contracts all the instants which separate us from it into a living present" (DR 77). This suggests that the condition of habitual time—futural time, difference-in-itself—is the wound. However, this transcendental privilege also becomes an ethical one when Deleuze insists on the importance of the time of apprenticeship and the way it never leaves us and is never fully mastered. His ethic is not one of phronesis, of practical wisdom within a given embodied and cultural context. If an ethics of phronesis can be seen tacitly at work in most phenomenology, with its communitarian inclinations, this is not the case for Deleuze. On the contrary, it is an ethic of jolting this world, and of disturbing the equanimity of the experiences of wisdom and mastery, just as Derrida likewise insists on the time of apprenticeship and the undecidability of friendship that is simultaneously a transcendental condition and also omnipresent as we will see in chapter 8. In relation to Deleuze's "secret dualism" which obtains between the transcendental (virtual) and the actual, it is worth noting that even in *Difference and Repetition* there is said to be a vast difference between "material and a spiritual" repetition (DR 84), between the actual and the virtual. And we have seen that even though the transcendental for Deleuze is not fixed but fluid (and in an asymmetrical relation of reciprocal determination with the actual), the virtual (via the quasi-cause) retains a priority over the body and states of affairs. It is an epiphenomenal and temporal wound that not only has a philosophical order of priority, but also an ethical one. As such, his work constitutes an ethics of the virtual, or an ethics of the event-effect. I hope to have shown that grasping this depends upon seeing the significance of his philosophy of time and its anti-presentism, along with the wound that time opens up both individually and virtually. I also hope to have shown that his ethical principles derive from a hierarchical transcendental philosophy that gives to the body the lesser role (even when Deleuze talks of sensations, it is important to note that they come from the virtual and the surface moreso than from the realm of bodies and depths). While transcendental "priority" ought to be understood neutrally by Deleuze, quite frequently something else is going on in his texts, which intermittently expresses itself (to greater or lesser extents), and which philosophers like Hallward, Badiou, and myself, have attempted to thematize, albeit in quite different ways. If one thinks this characterization is too swift given Deleuze's Spinozian declarations that we do not yet know what a body can do (and even here, again, such transformations can and must come from outside the body, from something akin to the virtual), certainly the mediating capacities and abilities of bodies are marginalized. It seems to me that despite the profound excavations that his work exerts upon the Cartesian mind, the philosophy of the subject, and the philosophy of representation, Deleuze's philosophy nonetheless reinvents a strange amalgam of the modern and the premodern, reinventing a form of dualism that is uniquely his own. No doubt that is a major accomplishment (understood in terms of the creation of concepts), but it is also one that, I think, deserves to have these and other critical questions put to it.

7

Deleuze on the "Perverse" Structure: Beyond the Other-Structure and the Struggle for Recognition[1]

Although it is not widely recognized by commentators on his work, Deleuze has some deep and interesting things to say about inter-subjectivity (and indeed the individual) that are worth examining in their own right, but also because they derive some normative judgments from claims to transcendental priority that are, once more, bound up with his distinctive philosophy of time. In *Difference and Repetition*, "Michel Tournier and the World without Others," *The Logic of Sense*, and "Coldness and Cruelty," Deleuze offers an account of our relations with others in which the normalizing structures thematized by versions of the struggle for recognition (and even the regulatory function of what he calls the other-structure itself), are seen as insufficient to account for aspects of desire and inter-subjectivity that he labels as part of the "perverse-structure," including, most notably, masochism. While Deleuze accomplishes his aim of showing that the phenomena of masochism resist being comprehended by any Hegelian-inspired theorization of the struggle for recognition, I will raise some questions about the extent to which his own alternative normativity is justified and suggest, again, that what we actually bear witness to is a constraint on transcendental philosophy and any project of grounding that contains a normative ethico-political impetus. While Deleuze's work highlights that the times of our lives are indeed fractured, I continue to argue here that he does not succeed in showing that the integrative and normalizing dimensions of embodied experience are either ethically insignificant and/or reactive coping in the face of a provocation and disruption that is its prior condition, in both a chronological and also transcendental sense.

Deleuze's Objections to the Centrality of the Struggle for Recognition

Let us begin by considereding Deleuze's relation to Hegel's famous descriptions of the master-slave dialectic, and the more general analysis of the struggle for recognition that it is a part of. Hegel's thematizations of this struggle for recognition in *Phenomenology of Spirit* have been remarkably influential throughout the nineteenth and twentieth centuries, being important to almost the entire Marxist tradition, Nietzsche, psychoanalysis (especially Jacques Lacan, Slavoj Žižek, and Donald Winnicott), existentialism (especially Sartre), feminism (Simone de Beauvoir, Jessica Benjamin, Judith Butler), the Frankfurt School, contemporary German theorists of recognition (especially Axel Honneth), and arguably also poststructuralism. In various different texts, however, Deleuze stridently distances himself from the oppositional logic at work in Hegel's conception of the dialectic of lordship and bondage, in which each of the interlocutors fights for, and depends upon, the recognition of the other, and in which self-identity is itself claimed to be founded through this dialectic. To present the key claim of this dialectic schematically, Hegel's account of the development of self-consciousness via a struggle for recognition, unlike say Thomas Hobbes' "state of nature" idea, puts the other at the heart of the self, as a condition of self-consciousness and the ability to reflect on oneself as an "I." Self-consciousness is produced through historical encounters with others, and we can understand this in many different senses, including developmental, phenomenological, historical, and even in relation to nation formation. In particular, the master's social identity is said to be thoroughly dependent upon the recognition of the slave, and, to an extent, the reverse also applies. However, at least according to Hegel's paradoxical analysis in *Phenomenology of Spirit*, there is a sense in which the slave is actually freer because their identity is not so bound up with the need for recognition from the master and because in their work they have an opportunity to transcend their socially designated situation. As is well-known, this idea was later taken up and reconfigured by Marx (in his contradiction-based class analyses) in order to explain the revolutionary potential of the proletariat who is not tied to bourgeois interests like property and recognition, or at least is not tied to them in a manner that is likely to perpetuate capitalism. Since then, other major European philosophers have also drawn heavily on the master-slave dialectic, without quite the same "end of history" narrative that has tended to accompany the Hegelo-Marxist versions. For Hegel, Marx, Honneth, and myriad figures in between, the opposition between the master and the slave, the landowner and the worker, is eventually supposed to be sublated and overcome, either by the movement of *Geist*, by material and structural transformation of the ownership of the means of production, or by a combination of the two. It is this teleological understanding of contradiction as both the cause of social transformation and the potential cure, that poststructuralism generally, and Deleuze in particular, rejects.

However, it is not only the "end of history" grand narrative, the delimitation of the future as both known and inevitable, that is called into question. Rather, as Deleuze makes clear, any priority given to the causal phenomena of opposition and contradiction (rather than paradox) also misconstrues difference by simplifying the complex of factors and problems that are at play (DR 50). In fact, he suggests that the appearance of contradiction—in the reified contraries of the master and the slave but also any other structurally equivalent opposition—is but an epiphenomenon, a derivative ossification of a more fundamental swarm of differences (a productive multiplicity) that is prior to, and resists, sterile class-stratified analyses of society.[2] He even provocatively tells us that contradiction is not the weapon of the proletariat but the manner in which the bourgeoisie defends itself, suggesting again its derivative status (DR 268). It is for similar reasons that in *Anti-Oedipus* Deleuze and Guattari object to the equally influential Freudian model of social relations, which focuses on familial contradiction (the opposing and impossible desires of the "mummy, daddy, me" triad) as the key factor in the channeling of desire and the determination of the psyche, and which excludes from consideration investments in the broader social milieu.[3] In *Difference and Repetition*, Deleuze also stridently states that the repetition in psychic life is dominated and undermined by the opposition at the heart of the theory of repression (DR 271), thereby intimating that, for all of its advances, psychoanalysis also ends up domesticating "difference-in-itself" despite the significant debts that Deleuze owes to it, notably in regard to the death instinct. Without being able to pursue the complicated issue of Deleuze's relation to psychoanalysis further, his enduring post-Althusserian insistence on a plurality of conflicts and interconnections, on an overdetermined multiplicity of causal influences, frees him from the grip of all of these reinventions of the master-slave dialectic that artificially and erroneously cut out a particular opposition from a larger milieu of overlapping perspectives.

We should note, however, that Deleuze does not simply seek to refute all versions of dialectical thinking. Despite the serious reservations about the dialectic that he expresses in *Nietzsche and Philosophy*, in other works Deleuze more typically refers to the long history of the *distortion* of the dialectic (DR 268). As the term distortion intimates, this does not entail a repudiation of the dialectic per se but a particular distorted version of it. In this respect his concerns are primarily with the posing of an opposition or contradiction between two forces (when there is really a multiplicity of forces), as well as the key role that is given to negation. And there is a priority accorded to the experience of negativity by theorists of the struggle for recognition. In particular, the other person is primarily apprehended negatively as a "not-me" who recognizes and apprehends a part of me that I cannot myself apprehend or control, and therefore alienates me from my transcendent projects in the world. This is certainly the case on Sartre's analysis, as well as on Hegel's and Alexandre Kojève's, where the experience of this negativity is considered vital. To summarise, then, it seems plausible to claim, as Deleuze does, that the various versions of the struggle for recognition privilege three fundamental tropes: contradiction, opposition, and negativity. Now these are clearly part of our experience of social life, but, according to Deleuze, they are not part of the fundamental level of desire and sociality. They pertain to the molar rather than molecular level to invoke the terms of *Anti-Oedipus*.

Other-Structure

Arguably one of the most ill-understood sections of Deleuze's oeuvre, both because it is rarely addressed and because of a general lack of agreement regarding how it fits in with his broader body of work, revolves around his enigmatic comments regarding what he calls the "other-structure," particularly in *Difference and Repetition* and in his essay "Michel Tournier and the World without Others," an appendix to *The Logic of Sense*. He also briefly addresses the issue in *Proust and Signs* (7-9) and *What Is Philosophy?* (17-9). In all of these books he points toward a new conception of the Other—the Other as expression of a possible world; as a structure that precedes any subsequent dialectical mediation and which hence avoids the theory of the master-slave dialectic of social relations as it has been construed since Hegel.

In *Difference and Repetition*, for example, he rejects the tendency of theories of social relations "to oscillate mistakenly and ceaselessly from a pole at which the Other is reduced to the status of object to a pole at which it assumes the status of subject" (DR 268). He targets Sartre for ultimately remaining within this paradigm, despite having seen that the Other is an irreducible structure of being. For Sartre, the ontology evinced by the mode of being-for-others is not reducible to the categories of either being-for-itself (the negations of consciousness) or being-in-itself (the plenitude of objects and things), but Deleuze argues that Sartre betrays this insight by positing the look as ontologically primary to, and constitutive of, the other-structure. Because of this, Deleuze suggests that Sartre cannot but return us to a looking (subject), looked at (object) model of social relations (DR 268). Deleuze says, even Sartre "was content to inscribe this oscillation in the Other as such, in showing that the Other becomes object when I become subject, and did not become subject unless I in turn became object. As a result, the structure of the Other, as well as its role in psychic systems, remained misunderstood"

(DR 260). But, for Deleuze, it is not just Sartre who presupposes such a theory of social relations, but also the phenomenological tradition more generally, arguably even to Levinas' "humanism of the other man."[4] In *What Is Philosophy?* Deleuze and Guattari implicitly accuse Levinas' account of the Other as reinstituting a model of "transcendence within immanence,"[5] but without dwelling on whether or not this is a fair assessment it is time to consider Deleuze's alternative understanding of the structure of the Other.

It is at once simple and difficult, and it is not a radical as one might expect. In fact, to a large degree Deleuze's initial move recapitulates the main thrust of Merleau-Ponty's critique of Sartre's view of social relations in *Phenomenology of Perception* and *The Visible and the Invisible*, where he describes Sartre's work as an untenable "agnosticism about the other" because it ignores the fundamental expressivity of the Other that precedes and makes possible all further developments, including those involved in the master-slave dialectic.[6] Deleuze's key claim is also that the Other "cannot be separated from the expressivity which constitutes it," from the frightened world that is given in the glimpse of another. The Other *is* its expressivity, the revelation given of a possible world (DR 260-1), and Deleuze reaffirms this perspective more than twenty years later in he and Guattari's analysis of the concept of the other person in *What Is Philosophy?*[7] Now, how we are to understand that expressivity that he refers to is no easy question, but it is clear that on Deleuze's view we must adhere to some particular conditions, conditions that are at least partly phenomenological ones. Indeed, Deleuze consistently refers to and intimates the need for a more radical reduction (DR 52, 137), and in this context he suggests that we must attend to "the moment at which the expressed has (for us) no existence apart from that which expresses it: the Other as the expression of a possible world" (DR 261). Only by performing this more radical reduction can we get beyond the apprehension of particular concrete others (and an associated commitment to the self or subject) to the a priori Other that is their transcendental condition. He insists on this distinction between concrete others and an a priori Other in all of the texts in question. He also argues that this a priori Other is best understood as a structure, and, more specifically, as the "structure of the possible."

What is meant by this formulation is the subject of some debate, not least because of the derogatory connotations that are usually attached to Deleuze's use of the term "possible." After all, Deleuze frequently criticizes philosophical positions that prioritize the possible for not attaining to what he considers to be the order of the true transcendental condition(s) of actual experience that he calls the virtual, which can be schematically understood as a transcendental realm of multiplicities and profligate differences that undermines fixed identities and is also generative of the new. As Deleuze suggests, "the possible and the virtual are distinguished . . . because the former refers back to the form of identity in the concept, while the latter designates a pure multiplicity in the Idea that radically excludes the identical" (DR 273). Nevertheless, it is clear that when referring to the Other as "structure of the possible," his use of the term "possible" does not refer to something abstract, or something that does not yet exist but might potentially do so. As he observes in his essay on Michel Tournier, the expressed possible world (e.g., frightened) "certainly exists, but it does not exist (actually) outside of that which expresses it," the final clause amounting to an insistence that we cannot understand the Other's terror simply on the basis of the empirical world that sets it off, or by analogy with our own experiences of terror (LS 346).[8] For Deleuze it is this other-structure that opens up both of these second-order possibilities, as well as possibilities more generally, including those given to us by the perceptual field and the experience of subjectivity itself. Rather than it needing to be explained how, given the fact of perception, one might manage to apprehend a particular concrete other, the other-structure is understood as the condition of possibility for perception. In a remark that is interesting, not least because it seems to imply that the other-structure is akin to the virtual, that is, a transcendental field of differences (an idea that he distances himself from shortly after), Deleuze suggests that:

> . . . the first effect of others is that around each object that I perceive, or each idea that I think, there is the organization of a marginal world, a mantle or background . . . an entire field of virtualities and potentialities which I already knew were capable of being actualized. Now such a knowledge or sentiment of marginal existence is possible only through other people (LS 344).

While phenomenology and gestalt psychology have made similar points, for our purposes the important idea to ascertain from this is that the "a priori Other," the other-as-structure, precedes and makes possible what Deleuze calls the relations of "development and explication" that often resemble something like the master-slave dialectic, as Sartre, Hegel, and others have described it, and in which one subject (tacitly or otherwise) attempts to be recognized by another. The expressivity that is the Other commands a response, and, for Deleuze, one has no choice but to "explicate or develop the world expressed by the other, either in order to participate in it or to deny it" (DR 260). Perhaps we develop the expressed possible world into a reality that consumes us; perhaps we denounce it as illusory. Here is where negation and opposition become part of the equation, but these relations of development dissolve the true structure of the Other. The expressivity of the Other as possible world (noting that this conception of the Other is not reducible to humans) precedes the kind of oppositional and contradictory permutations that social relations typically devolve into, even if that is not to say that they are necessarily unhappy or hell-like as Sartre had his character Garcin dramatically observe in *No Exit.*[9]

From describing these transcendental and quasi-phenomenological conditions of social life, however, Deleuze rather quickly derives an intriguing ethico-political injunction: not to explicate oneself too much with the Other, and not to explicate the Other too much, but to "multiply one's own world by populating it with all those expressed that do not exist apart from their expressions" (DR 260). This implied rejection of explication follows philosophically from the priority that he accords to the expressivity of the Other, because understanding social life from that perspective of explication misconstrues the structure of the Other in a manner akin to remaining within the natural attitude. Nevertheless, the moral resonance of this comment gives us reason to pause. On what basis does Deleuze derive his injunction to multiply these possible worlds, these a priori expressed others that have not yet been explicated or thematized? It seems that the transcendental condition (the other-structure as expressive of a possible world) is simultaneously a moral injunction to maximize actual occurrences of such expressivity. While it is not completely clear how to understand this general principle that is always under the surface in Deleuze's work, and that James Williams summarizes as the imperative to maximize both forgetting and connecting,[10] we can begin to tease this out by considering Deleuze's comments on love that immediately follow his discussion of the expressivity of the Other. Deleuze suggests that, "there is no love which does not begin with the revelation of a possible world as such, enwound in the other which expresses it" (DR 261). In this sense, the expressivity of the Other would be that revelation of a possible world, that event from which all that is beautiful and bewildering began, and which will thereafter be transformed by relations of development and explication—the other person as fascinating subject one minute, deceptive object another. But there are several questions that need to be raised here. Is the implied denunciation of relations of development and explication justified? After all, while relations of explication might come to domesticate the Other's "otherness" and to partially deprive them of their radical difference, as Deleuze suggests, it is also the case that they open up different and more diverse kinds of relations that cannot be captured on this view that juxtaposes (and privileges) the relative purity of expressed "possible worlds" that have no ties of allegiance (i.e., the different and the new), against their shutting down and increased monotony in the world of identities. Even if a transcendental priority is accorded to difference and the new, it is not morally clear that we can enrich the world only by promulgating encounters with these uncontaminated "possible worlds," and while Deleuze acknowledges that any "possible world" must end up being actualized, he cannot but appear to lament that this is so (we will see that this is particularly apparent in his essay on Tournier). To put the problem another way, even if the condition for relations of explication (a quarrel, a revelation, anything that remains with the play of identities) is the other as possible world, it does not necessarily follow from this that we should live privileging this transcendental condition, or even the intensities and singularities that this condition actualizes. Indeed, while Deleuze himself repeatedly insists that there is reciprocal if asymmetrical determination between the actual and the virtual, which means that neither legislates and draws up limits or rules for the other (whereby we might, for example, obtain clear moral rules about what should take place in the actual), in practice the virtual (i.e., difference, the play of surfaces) clearly plays the determinative role in his injunction to multiply encounters with the expressivity of others.

If we think about this practically for a moment, and no doubt more prescriptively than Deleuze would himself countenance, what does this advocation of multiple encounters without explication and development commit Deleuze to? To pose the problem bluntly, is Deleuze committed to advocating a multiplicity of encounters and the radically new expressions of possible worlds that are opened up prior to imbrication and explication? Or, is he tacitly committed to what we might call perverse impulses of fidelity to particular kinds of moments of expressivity of the Other, modes of living that have long held his attention in highly acute studies of sadism, masochism, schizophrenia, etc? "Coldness and Cruelty," and the expressions of fear and trembling are, after all, as revelatory of a possible world as anything else, and in *Difference and Repetition* Deleuze explicitly links the desexualization and then resexualization of the third synthesis of time (the privileged time of the future) with sadism and masochism (DR 114-5). While the rejoinder to this suggestion might be that this more fetishistic option does not have enough forgetting involved, to recall Williams' maxim, it can be shown that there is often something close to an endorsement of masochism in Deleuze's work and the general transcendental priority that he also gives to what he calls the "perverse-structure" will be detailed shortly. It seems to me, however, that Deleuze is committed to the one or the other of these alternatives, and this is largely because processes of embodied comportment to one's situation (habit and skilful coping within a view to maintaining something like an equilibrium) are tacitly diminished in his work, remaining radically disjunct from the realm of genuine learning, thought, creativity, the virtual, and that which is of value in the other.

If this is not already problematic, later on in *Difference and Repetition* Deleuze also briefly entertains going beyond the apprehension of the Other as noone (i.e., the a priori other that is the structure of the possible), to:

> where we reach the regions where the other-structure no longer functions,
> far from the objects and subjects that it conditions, where singularities are

free to be deployed or distributed within pure Ideas, and individuating factors to be distributed in intensity. In this sense, it is indeed true that the
thinker is necessarily solitary and solipsistic. (DR 282)

Here then, we see a clear reference to the virtual as being beyond, or prior to,
the other-structure (the other as expressive of possible world), thus troubling
James Williams' interpretation of the other-structure as synonymous with
the virtual,[11] but also troubling, in a slightly different manner, Constantin
Boundas' interpretation.[12] While this comment is not explored in any detail
in *Difference and Repetition*, it is developed at length in Deleuze's almost
contemporaneous discussion of Tournier's novel, *Friday*, and it hence cannot
be dismissed as an inconsequential aberration. We will examine Deleuze's
text on Tournier shortly, but it is worth foregrounding some of the difficulties with this position. Firstly, and immanent to Deleuze's work given his
new conception of the Other (as expressive of a possible world, difference,
etc.), does he have a need within his philosophical system to "go beyond"
the other-structure? And what might this going beyond consist in? Who are
these thinkers that can create without an other-structure? Isn't it Deleuze who
has shown us that something always provokes us to think? What could that
provocation be, if not the Other, here understood as synonymous with problems? From my own point of view, while Deleuze offers us an illuminating
account of the other-structure the rather romantic normative position that he
derives from it is more troubling.

Michel Tournier and the World without Others: Beyond the Other-Structure

This ethic becomes most obvious in his intriguing essay "Michel Tournier
and the World without Others", which is oriented around Tournier's book,
Friday, which is itself a rewriting of the Robinson Crusoe narrative. In this
essay, Deleuze develops this theme of the other-structure, as well as what his
mooted "going beyond" of the other-structure might entail, but it must be
noted that the other-structure takes on a far more negative understanding than
we have seen in *Difference and Repetition*. On Deleuze's reading, Tournier's
tale suggests that Crusoe, finding himself alone and despairing on the island,
first strengthens and reinstates an other-structure by neurotically setting up an
order of laws, rules, and habits (skills and habits are again treated as synonymous with rules and norms, as they are in *Difference and Repetition*). In this
context, Deleuze also suggests that this other-structure ensures "the sweetness of contiguities and resemblances," "fills the world with a benevolent
murmuring," and provides for "transition" (LS 345). Later on he suggests,

"the other-structure organizes and pacifies depth. It renders it liveable" (LS 353). This view of the other-structure no longer seems to be so clearly associated with the expressivity of a possible world. On the contrary, his comments about it overlap and dovetail with his contemporaneous critique of much of Western philosophy for presupposing an ur-doxic harmony between self and world, and self and others. The references to depth are equally important, suggesting that the other-structure remains within the realm of depth (as do bodies and chronos, the time of the present), not yet attaining to the play of surfaces which, in the main text of *The Logic of Sense*, it becomes clear is the pure event (the incorporeal time of Aion), the true transcendental that is synonymous with what he later calls the virtual (LS 126, cf. 61, 72, 105).[13] Similarly, it is notable that both contiguities and resemblances, the terms used to describe the other-structure in the Tournier essay, are given a thorough problematization in both *Difference and Repetition* and *The Logic of Sense* (LS 123, 141, 165). But does this implied rejection of the philosophical priority of this world-giving aspect of the other-structure, we might say its "generosity" invoking the work of Rosalyn Diprose,[14] follow? And even if it does, does it legitimize the subsequent morality?

This is an important issue. Merleau-Ponty's philosophy, for example, whether in *Phenomenology of Perception* or *The Visible and the Invisible*, might well be said to agree with Deleuze's descriptions of the other-structure but argue that it does not commit one to harmony and that there is no going beyond this chiasmic intertwining. Perhaps we are returned here to a distinction between Merleau-Ponty's lived phenomenology and Deleuze's apparent desire for a phenomenology without reference to the lived, which evince two different kinds of pragmatism. Merleau-Ponty's ontology never quite leaves phenomena and the body behind. Even *The Visible and the Invisible* is still based on certain phenomena (i.e., the famous example of two hands touching, as well as other chiasms that are phenomenologically evident). For Deleuze, however, metaphysics leaves this "lived" behind in the importance that he accords to the virtual, which tacitly legislates for the actual in an asymmetrical move that sometimes seems to simplify what is involved in our embodied comportment to a situation. Deleuze wants to ascertain transcendental conditions for it and privilege those conditions that lead to creativity; Merleau-Ponty wants to derive transcendental conditions from the body, although these are generalizable far beyond the realm of the human (i.e., the flesh). Merleau-Ponty's transcendental philosophy of normality is thus perhaps opposed to Deleuze's self-proclaimed transcendental philosophy of the perverse. In this light, we can understand Foucault's observation that *The Logic of Sense* is the most alien book imaginable compared to *Phenomenology of Perception*,[15] but my wager, against Foucault no doubt, is that we are best served seeing the imbrication of these accounts rather than privileging the one over the other as the philosophy of the twenty-first century.

But to return to the Tournier narrative, after various neurotic tribulations Crusoe is portrayed as enacting a kind of ontological shift that leaves behind the other-structure (and therefore also the structure of possibility), such that even when Friday eventually arrives on the scene he is not encountered as another person. Friday, Deleuze tells us, is not an Other, but "something wholly other than the Other," who "dissolves objects, bodies, and the earth" (LS 355). Without this structure, the category of the possible collapses and "gone is the sweetness of contiguities and resemblances which allowed us to inhabit the world. Nothing subsists but insuperable depths, absolute distances and differences, or, on the contrary, unbearable repetitions" (LS 345-6). After a period of time, Deleuze suggests that rather than lament the loss of the other-structure, for Crusoe it is this structure itself that begins to be seen as the problem, as the progenitor of all of these invasive possible worlds (LS 350).[16] Moreover, once he is able to move beyond the other-structure, Crusoe clings to objects and has a "generalized erection"—that is, an erection of surfaces. The pure surface, Deleuze observes, is perhaps what the other-structure was hiding from us, and it should be noted that this links in with the transcendental privilege that Deleuze accords to both the surface and the time of the event (Aion) in *The Logic of Sense*. On this level of the surface, the depth of the other-structure is seen as "imprisoning elements within the limits of bodies," and Deleuze is "tempted to conclude that bodies are but detours to the attainment of images" (LS 351-2). Deleuze even asks, "when we desire others, are not our desires brought to bear upon this expressed possible world which the Other wrongly envelops, instead of allowing it to float and fly above the world, developed into a glorious double?" (LS 354). He intimates that perhaps, "the absence of the Other and the dissolution of its structure do not simply disorganize the world, but, on the contrary, open up a possibility of salvation" (LS 354). Perhaps Deleuze might respond that he is here merely voicing the logics of Crusoe's perversion and analyzing Tournier's novel rather than making a philosophical statement. Such a response is, however, disingenuous, despite the fact that Deleuze flirts with it himself—"everything here," he says at one stage, "is fictitious" (LS 356). After all, we have seen that in *Difference and Repetition* he also refers to a leaving behind of the other-structure, and in his essay on Tournier the other-structure (envisaged as organizing and regulatory) is consistently shown to be disrupted by the perverse-structure that is so evocatively captured by Tournier's descriptions of Crusoe, and by the many and varied other literary authors and characters that Deleuze has been concerned with (think of Proust, Fitzgerald, Beckett, etc.).

In fact, it is arguable that all of Deleuze's work enumerates a metaphysics of what he himself calls the perverse-structure, which is the condition of actual perversions and which is also opposed to the other-structure. Would

it be remiss to suggest that his philosophy of difference hence itself constitutes a world without others, a perverse-structure that attempts to dispel the
normativity inherent in the other-structure? After all, we have seen that he
explicitly discusses the solipsism inherent in creativity, directly links the solipsism of sadism and masochism with the privileged third synthesis of time
in *Difference and Repetition*, and in his final collaborative work with Guattari
stridently rejects any privileging of communication.[17] Of course, there are
compelling reasons for all of these moves, and his target in the latter respect
is more a particular philosophy of communication (i.e., Habermas') rather
than communication per se, but it is nevertheless clear that such a world without others would not be the lived world. And it is also worth noting, contrary
to a certain reception of his work, that Deleuze is consistently wary about the
value of the lived (and the lived body), generally accusing any philosophical
explication of it, and certainly any philosophical reliance upon it, of depending on common sense and good sense and thus as ultimately amounting to a
conservatism—this is most clearly shown in *What Is Philosophy?* and *The
Logic of Sense* (LS 346).[18] While Deleuze continually insists upon the importance of the idea of reciprocal determination between the actual and the virtual, it is from the virtual (or from the transcendental topology of surfaces in the
terms of *The Logic of Sense*) that he derives the morality that is bequeathed
to his descriptions of the other-structure, both in the injunction to multiply
expressions of possible worlds (which constitutes the most obvious ethic of
Difference and Repetition) and in the injunction to leave such worlds behind
(which develops a thread from *Difference and Repetition* and constitutes the
ethic of his Tournier essay). This is problematic on Deleuze's own terms, but,
even without considering the morality that is sometimes betrothed to his philosophy of difference, it is not clear just how these transcendental conditions
(differences, repetitions, a transcendental field without an other-structure)
can do without the transcendental condition of the other-structure that is, as
Deleuze himself admits in multiple places as we have seen, a priori. While
Deleuze's non-Kantian conception of a priori conditions (i.e., as a variegated
and nonuniversal transcendental field that is in a reciprocal relationship with
the actual) means that a priori conditions can be changed and transformed,
important questions remain about how these different transcendental levels
might interact that are never adequately thematized in Deleuze's sometimes
contradictory writings on the other-structure. As we have seen, it is sometimes treated as of the order of the virtual, sometimes like the event, but
more often as merely of the order of actuality and depth. The question of the
other-structure is hence, it seems to me, an aporetic blindspot in Deleuze's
work, around which many key issues in his philosophy are bound up and co-
implicated.

Masochism, Perversity, and the Other-Structure

While Deleuze embraces the perverse structure as the condition of the other structure in his Tournier essay, he makes a related move in his various reflections on masochism, including in "Coldness and Cruelty," "From Sacher-Masoch to Masochism," and *Difference and Repetition*. His aim in such texts is to firstly offer an adequate symptomatology of masochism, and he maintains that if this is done it will be clear that the master-slave dialectic (or structurally isomorphic correlates, like the positing of a single sadomasochistic entity) not only does not understand these ways of existing, but necessarily could not. If we are attentive to the literature of Leopold von Sacher-Masoch, progenitor of the term masochism, as well as to a lesser extent the literature of Marquis de Sade, progenitor of the term sadism, Deleuze contends that we will see that the idea of a single "sadomasochistic entity" is a crude simplification, a badly analyzed composite of symptoms that is reliant upon hasty causal assumptions. It issues forth from a confusion of syndromes with the specific symptoms involved in the two kinds of behavior.[19] While Freud and Richard Krafft-Ebbing and much of the medical profession repeatedly linked the two causally, positing a single sadomasochistic syndrome, literature can show us their radical differences by isolating particular ways of existing, and by giving us what we might call a more radical phenomenology that allows the differentiation of true symptoms from false syndromes that generalize. For Deleuze, any adequate symptomatology will show us that sadism and masochism are not adequately understood as opposing but mutually reinforcing modalities of being. I will suggest in what follows that Deleuze is right to suggest that there are different and mutually exclusive logics and symptoms involved in sadism and masochism, differences that have tended to be obscured partly because of the influence of the master-slave dialectic and the correlative positing of a single sadomasochistic entity, but I will argue that the transcendental devaluation of the latter (as akin to coping) is not justified, and that we are, once more, better served seeing the coimbrication and irreducibility of these two forms of desire.

But first, let us briefly attend to the connection between the master-slave dialectic and the positing of a sadomasochistic entity in Sartre's phenomenology. After all, if the master-slave dialectic (and the more general struggle for recognition) is a pivotal part of social life, and functions through negation, opposition, and contradiction, then one would expect that sadism and masochism would be diverging responses to intersubjective life that would find in their opposite at least a temporary solution to their respective needs and desires. Indeed, this is precisely what Sartre famously argues in his bleak portrait of intersubjectivity in *Being and Nothingness*. In his view, the relationship between sadism and masochism is basically equivalent with the structure of Hegel's master-slave dialectic, except that Sartre, unlike Hegel, sees no possibility of overcoming this situation.[20] For Sartre, we are envisaged to be perennially thrown back and forth between the attitudes of sadism and masochism in an ultimately impossible attempt to control how we are seen by others, and to eliminate the prospect of shame and alienation before the look of the other. All forms of desire and love are understood on this model, and Sartre arrives at this rather pessimistic conclusion on account of the fundamental role that he gives to the phenomenological experience of shame in the looker/looked-upon dyad, on which he bases his analyses of concrete human relations.

Sartre famously describes a person peering through a keyhole into the next room, entirely absorbed in their activity. Suddenly though, they hear footsteps in the corridor behind them and they are aware that somebody is now watching them. No longer concerned with what is going on behind the door, they are aware only that they are the object of another's look and that they are being evaluated and judged in ways that they cannot control. They are reduced to an object in that other person's perceptual field, and this is the original meaning of relations with others.[21] As such, the other is primarily apprehended *negatively*, as a "not-me" who recognizes and apprehends a part of me that I cannot myself apprehend. Michéle Le Doeuff calls this Sartre's "de facto solipsism."[22] Although there are certain variations within the alternative roles that people might adopt as a response to the phenomenological feeling of shame, for Sartre one is essentially either the looker or the looked-at, and he insists that two people cannot simultaneously look at each other in his ontological sense. On his dialectic of social relations, the existence of a looker presupposes an interlocutor who is the looked-at, and sadism and masochism are two opposing ways of dealing with this dilemma—we can constantly judge and objectify others and thereby seek to prevent the emergence of our social self, or we can try and induce others to see us exactly as we wish to be seen and thereby control their subjectivity. More to the point, a masochistic project requires a sadistic collaborator, and vice versa, even though on Sartre's view both projects are doomed to failure since neither can be stably maintained.

While empirical data and the literature of Sacher-Masoch and de Sade arguably undermines this analysis, since sadism and masochism seem to be more stable and enduring than Sartre's account suggests, the key problem with it seems to be that certain assumptions simplify the mode of being of the masochist (and perhaps also the sadist, although that is nor our prime concern). Firstly, all is interpreted through the lens of this alienating look and an ensnared consciousness suddenly without time and history. When any particular background is given, however, it becomes clearer, Deleuze maintains, that masochists do not seek merely to be seen as an obscene object as

Sartre's analysis suggests, in order to mitigate against the possibility of being surprised by how the other views us and/or divest themselves of all subjectivity because of the anguish it induces. For Sacher-Masoch and Deleuze, masochism is a more complicated project than that, one that retains subjectivity and activity in the manner in which their accomplice is seduced into being a "quasi-sadist" (this term "quasi" is very important as we will see), and it is this process, along with the associated rituals, that is pleasurable. Indeed, Sartre's analysis gives little attention to the sense in which masochism constitutes an attempt to subvert, or play with, the typical relationship between pleasure and the law, something that Deleuze's more psychoanalytic account develops. Moreover, Sartre's explanation is unable to account for Sacher-Masoch's explicit desire for a third party to intervene between him and the woman he loved (and their contractual arrangements), precisely because part of what is at stake, for Sacher-Masoch, is to show how the contract that attempts to preclude this eventuality is necessarily undermined. On Sartre's analysis, masochism and sadism seek to exclude the third and shore up a dyadic relationship,[23] and he hence underestimates the performative dimension of masochism, the way in which a law is set up precisely for it to be problematized and turned against itself over a long period of time.

Let us consider, then, some of the main differences, albeit nonoppositional ones, that Deleuze highlights between these two typologies, sadism and masochism. In *Difference and Repetition*, Deleuze tells us that sadism functions by ascending to principles, but principles understood as some kind of original force, whereas masochism descends toward consequences to which one submits with all-too-perfect attention to detail, and it tends to involve demonstration by absurdity and working to rule (DR 5). In "Coldness and Cruelty," where this difference is given far more prolonged attention, sadism is said to focus on the institutions that render the law unnecessary and even obsolete. Replaced by a dynamic model of action and authority, sadism seeks the degradation of all laws and the establishment of a superior power. But, for Deleuze and de Sade alike, the impetus behind sadism is not simply the desire for power over others. Rather, it seeks to suspend what Deleuze calls the other-structure itself. The key aspect of sadism consists in the idea that the law can be best transcended through a kind of institutional anarchy that ascends to reasoned principles, but reasons and principles that somehow exceed and promulgate themselves, and that thus question our everyday normativity. An idea is taken to extremes, compulsively repeated. By contrast, Deleuze tells us that masochism highlights the way in which it is the contract, or agreement (tacit or otherwise), between parties and people that generates the law, before then focusing in detail on the inevitability of the way in which the subsequent development of the law then ignores or contravenes the very declaration that brought it into being. For him, these are very different ways of treating and overturning the law. Rather than rely on the moral law of convention, sadism surges upward to find rationality, living its own life, devoid of reference to custom, but masochism immanently shows the unjustifiable severity of law in the performative enaction of it.

More to the point, they instantiate different ways of responding to the relationship between law (including social norms) and pleasure, which are far from complementary opposites; the fantasy of the masochist is not, for example, predominantly about a sadist inflicting pain upon them. On the contrary, according to Deleuze's analysis of Sacher-Masoch's work, the masochist may not even find pain pleasurable. It is more likely that the experience of pain is a precondition of pleasure, not the same as it, and that the intersubjective seductions and anticipations also offer a different kind of pleasure. Moreover, unlike the sadist who sees the law as needing to be destroyed for pleasure, the masochist finds pleasure in the performative dimensions of the law. They join the law in a sense, but surreptitiously subvert it from within. Both sadism and masochism, on Deleuze's account, are a response to patriarchal law: the former seeks pleasure in the abolition of law (pleasure and law are viewed as antithetical); the latter seeks pleasure in the law. Likewise, the attitude to fantasy is also very different. As Deleuze observes, the masochist needs to believe that they are dreaming even when they are not, but sadism needs to be actual, to believe that they are not dreaming even when they are.[24] It is hard to see how these very different attitudes toward the law (including social norms) and pleasure can coexist in any given dyad, or the sense in which a causal and psychological connection might obtain between them in any single psyche, although we cannot consider the Freudian account in any detail here.

These are not peripheral interests to Deleuze's philosophy. In the introduction to *Difference and Repetition*, arguably his most important book, Deleuze insists that if the repetition of difference is possible, it is as much opposed to moral law as it is to natural law, and he then briefly discusses what he considers to be the two major ways of overturning the kind of repetition of the same that he associates with the moral law—sadism and masochism (DR 5). In sadism and masochism, repetition is said to run wild, of its own accord, and is no longer related to experience or to the pleasure that is gained or anticipated to be gained. Repetition is sought for its own sake and this can be problematic (i.e., it can lead to pathological fetishes), but, for Deleuze, again, the meaning of these terms cannot be confined to this. Rather, sadism and masochism are understood as the affective dimension of a certain transcendental structure and privileged synthesis of time that he calls repetition-for-itself (i.e., repetition that is not tied to identity). There are some important connections between his analysis of the *repetition-for-itself* evinced in sadism and masochism, and his understanding of the *difference-in-itself* at stake in the futural synthesis of time that he associates with the eternal return of difference, and what he terms the "apprenticeship of learning," experiences where one is radically and traumatically disrupted, forced to instigate new ways of existing to cope with difference and adversity (difference hence comes first, rather than embodied coping techniques). If sadism and masochism have this transcendental import that Deleuze intimates, if they capture one aspect of the synthesis of time and experience that is at work in all of us, as seems to be his implication in *Difference and Repetition*, this installs time and affect at the heart of politics: there can be no politics that is not bound up in a lived response to the problem of repetition.

The relation to time involved in sadism and masochism is also markedly distinct. Although both aim, according to Deleuze, to suspend the time of the living-present and open on to the time of Aion—the time of the eternal and the event—sadism and masochism do this very differently; they involve a respective acceleration and deceleration of time.[25] Things speed up with the calculations of time and the sadistic expansion of principles beyond law; the living-present becomes so compressed and hurried as to be obliterated. On the other hand, masochism is about a certain experience of waiting that tries not to anticipate or circumscribe the future by weighing it down with the expectations that are built into the habitual present, or what Heidegger called pragmatic temporality and which we examine in chapter 9. As we see detailed in Sacher-Masoch's novels, both the seduction and the rituals involved may be insidiously slow, allowing a relationship to slowly transmogrify and allowing the depth of one's coimplication with their interlocutor, who can never be an unequivocal master, to build and build. In this respect, it is no coincidence that Deleuze is also very enamored with the literature of Beckett and Proust, who, although very different, can both still be said to instantiate a different and more masochistic relation to time; the one prioritizing waiting and the other a memorial ritualism.

Upon consideration of the work of Sacher-Masoch and de Sade, then, Deleuze argues that we are "struck by the impossibility of any encounter between a sadist and a masochist."[26] Not only are they different modes of being with differing logics, but he also insists, contrary to Freud, that the existence of a person who is a masochist, for example (and the reverse also applies), does not imply the existence of an antagonistic sadist who inflicts suffering upon the masochist. Deleuze argues that a genuine sadist would never tolerate a willing masochist accomplice, and the whole point of masochism on his analysis is that any so-called punisher must first be educated and seduced into behaving in a manner that he terms "quasi-sadistic," which he rigorously distinguishes from sadism proper. They are more akin to separate ways of life that admit of no such clear-cut oppositionality, and the perceived failure of one of these two attitudes does not motivate us, as Sartre suggests in

Being and Nothingness and Freud suggests in his work both early and late, to adopt the alternative perspective.[27] Sadism, and more particularly masochism, hence challenge the master-slave dialectic and loom as something that it cannot adequately explain without simplifying or falsifying what is involved, most notably by the positing of a sadomasochistic entity.

But is it not possible that these aspects of social and sexual life might be accounted for by a new and improved version of Hegel's master-slave dialectic? Might it not be said that the inability to achieve mutual recognition often leads to a totalizing desire for domination as Jessica Benjamin maintains, and hence sadism and masochism can be understood on this model? Indeed, the move to absolute negation evinced by the sadist, and even the desire for submission found in O in *The Story of O* are seen by her to "represent a particular transposition of the desire for recognition."[28] The position of the sadist does seem able to be accommodated within the terms of the Hegelian system and its psychoanalytic developments. Likewise, Benjamin's Hegelian-inspired analysis of O does make perspicuous certain features regarding why one might consent to, and even wish for, submission to a powerful other.

Deleuze, however, would maintain that something more subtle is at stake with masochism proper, as opposed to the quasi-masochism he would associate with O. Recognition and social norms (law and the relation to pleasure) are precisely what is being played with in masochism, and the desire for masochistic relationships is not reducible to a desire for submission. In fact, it is the whole active-passive, powerful-impotent binary that is disrupted. Now, this is not to say that masochism is entirely outside of the desire for recognition, but that this desire is bound up with other desires (along the lines of the Freudian death drive) that are not so easily understood in terms of a battle for recognition, and that it also plays with this desire for recognition and its putative primacy. Indeed, this account of the death instinct (and associated desires, like those in masochism) as central to life rather than opposed to it, is very important to Deleuze in differentiating his view from those indebted to the master-slave dialectic, since there is a sense in which the Hegelian, Sartrean and Kojèvean positions all oppose life and death: self-consciousness is made possible by a formative encounter with the threat of death (a restricted conception of death), and the significance of life and its relation to freedom is secured through this opposition.[29] This opposition between life and death is perhaps not so stable, however. What if there is a death instinct that needs to be understood as something that is part of life, rather than something that is opposed to life, as Deleuze's appropriation of Freud maintains? These are questions that Hegel's account of desire and social life does not really consider, and they are questions we will return to in the course of examining Derrida's engagement with Heidegger on death in chapter 12.

Of course, even if it were accepted that masochism testifies to a form of desire (and intersubjective life) that the Hegelian story does not have the resources to comprehend, it might still be responded that sadism and masochism are but marginal cases, the exception that proves the rule, which is the dialectic of recognition enumerated in so many different ways by philosophers over the last couple of hundred years. Deleuze, again, would disagree. The formal characteristics of both sadism and masochism are seen to be present, to greater and lesser extents, in all of us, and I think he is right about this. These two affective (and logical) tendencies cannot be thought of simply as pathological activities on the margin of normality, and, in the case of masochism at least, they involve a movement of desire and a structure of interpersonal relations that is not reducible to those thematized by the master-slave dialectic (the other-structure). This does not, however, entail a direct repudiation of the master-slave dialectic or the struggle for recognition. After all, such a view clearly captures some of the constitutive conflicts in social life and it explains a movement toward normalization and integration that is bound up with communicative action. As Habermas and Honneth would maintain, moreover, many ethical norms do depend upon reciprocal recognition of self-conscious agency and identity. Following Deleuze in *The Logic of Sense*, we might say that the struggle for recognition offers an important theoretical elaboration of the other-structure, which is organizing and regulatory. On the other hand, Deleuze's philosophy and the literature of Sacher-Masoch and de Sade highlight the significance of something more akin to what Deleuze calls the perverse-structure, which subsists beneath "the sweetness of contiguities and resemblances which allowed us to inhabit the world."

For Deleuze, however, the perverse-structure is a condition for the other-structure (as the molecular is a condition for the molar), and he allies a morality with the perverse-structure. Deleuze's claim is hence more than merely that the other-structure and the perverse-structure are engaged in an ongoing dialectic in which they interrupt each other without any possibility of synthesis. Rather, he describes this kind of perverse-structure as the transcendental condition for the other-structure, and, he implies, as superior to it. In other words, there is a metaphysical or ontological hierarchy at work here. In-itself, it might be suggested that this is unproblematic, but we have seen that this hierarchy, which is perhaps most noticeable in his account of the other-structure, repeatedly also becomes a quasi-morality. The perverse is prioritized, but precisely why is never legitimated, other than through reference to its role in the genesis of creativity. We have seen his quasi-moral call for expressive encounters with others without relations of explication, and we have seen his inconsistent and somewhat mystical desire to move beyond the Other as expressive of possible world to the perverse-structure that is its condition. In this respect, I want to distance myself from Deleuze's transcendental philosophy that remains speculative, and instead maintain that these movements are equiprimordial. Both desires are present in us: a desire for the stable and law-like that makes possible communicative norms, as well as a subversive desire. To ignore the perverse-structure, or to attempt to explain it within a form of dialectical thinking that is ensnared by contradiction, opposition, and negation, is to perpetuate what Merleau-Ponty calls a bad dialectic. Likewise, however, to privilege the "perverse" over the normative through a sometimes illegitimate use of transcendental arguments in which a neutral order of transcendental priority also surreptitiously becomes an ethico-political one, is to risk lapsing into dogmatism. We have seen that both the sadist and the masochist differently seek to overturn the law, and to undermine the habitual movement toward an equilibrium on which it is suggested the law is based. But whether or not habitual coping and the law are reducible to one another, however, hence becomes a key question and it seems to me that such a conflation is problematic and illegitimately either passes over the complexities of the former (embodied coping) in tacitly devaluing the latter (law, norms and convention). Do habitual and embodied coping techniques merely cover over these more fundamental dimensions of time and repetition, as Deleuze's own analyses in *Difference and Repetition* suggest? And even if they do, ought we to live in accord with this transcendental dimension? Never adequately answering these kind of questions, it seems to me that Deleuze's work, as important as it is for showing the indispensability of time to subjectivity, is afflicted by its own chronopathology.

8

Derrida, Friendship, and the
Transcendental Priority of the "Untimely"[1]

Despite the intertwining that Derrida insists obtains between time and space, it is difficult to dispute that vast aspects of his work, both early and late, have been preeminently concerned with time. The strategy of deconstruction, of course, borrows from Heidegger's critique of Western philosophy's metaphysics of presence: for Heidegger, our philosophical tradition (and culture) has been unable to understand time except as just another present entity, albeit perhaps of a special kind. While Derrida challenges Heidegger's own purported overcoming of this "vulgar" and metaphysical treatment of time, suggesting in "*Ousia* and *Grammē*" that there is no other concept of time that might be opposed to this metaphysical one, it remains the case that his work consistently invokes a time (albeit an "untimely time") that disrupts presentist time. Moreover, as has been widely commented on, this disruptive thought is often associated with the future, which haunts the time of the present and cannot itself be rendered present. In this particular chapter, however, I wish to examine the way in which Derrida's account of friendship in *Politics of Friendship* deepens his abiding deconstructive insistence on the *contretemps* that breaks open time but nonetheless pertains to it. Although I will suggest that Derrida's temporal (and transcendental) deconstruction of the Aristotelian distinction between utility and perfect friendships is convincing, I will argue that his reservations about any "omni-temporal" account of friendship, which in a quasi-Aristotelian fashion emphasizes the centrality of stable dispositions and embodied *phronesis*, is less successful. Derrida's objections to an understanding of friendship based upon such phenomena revolve around the way in which their recuperative and binding component, based on habitual syntheses of time (e.g., the scar), covers over a certain structure of time that is considered to be both ontologically prior and ethico-politically more important (e.g., the wound of time: the immemorial past that nonetheless subsists; the future that defies our expectations and is the condition for the event). However, in the manner in which Derrida's position slides

147

between (1) being an argument for this aspect of time as a transcendental condition, and (2) nonetheless constituting an ethical imperative of sorts, I will argue that Derrida's philosophy is itself touched by time in the sense of "touched" that connotes affected and wounded. Indeed, Derrida's work appears to instantiate what Husserl might call a transcendental pathology, in that it intermittently instantiates an ethics of nonpresentist time (the time which is simultaneously the transcendental condition for the event), and, by contrast, disparages what we might call an ethics of *phronesis*, a lived friendship of "omni-temporal" dispositions, of embodied and habitual patterns. As well as showing this tendency in his work, I end this chapter in a manner that parallels the last two chapters on Deleuze: that is, by proposing a dialectic between these disjunctive and the conjunctive aspects of time (wound and scar) that does not accord any kind of a priori privilege to the one over the other.

Untimely Wounds, Aporias, and the (Im)possible Event

Before turning to the texts and themes that will be our main concern, however, a few contextualizing remarks are required concerning the motif of the wound that I have employed in the introduction to this chapter because it recurs frequently in Derrida's work, just as it also does in Deleuze's as we have seen. There are at least two ways in which Derrida uses the term wound. Firstly, to intimate the separation that obtains between the orders of the possible and the impossible, in his preoccupation with what has come to be termed "possible-impossible" aporias—that is, with themes in which the condition of their possibility is also, and at once, the condition of their impossibility. Derrida's paradoxical analyses of the gift, forgiveness, hospitality, etc.[2] all obey this logic in which, for example, the aporetic necessity to forgive the unforgivable constitutes a wound that is not susceptible to scarification. In this respect, Derrida speaks of the aporia that obtains between two terms of this kind as itself a wound: "tragically irreconcilable and forever wounding" (PF 22). This is the wound of disjunction between two laws that will not gather, exemplified for Derrida by democracy, which paradoxically seeks to respect singularity (and minorities) but also simply calculates majorities by voting and wherein each person's vote is theoretically substitutable or exchangeable for any other. As we will see, this refusal to gather, to bind, is very important to his work. It means that a theoretical resolution is not possible, and that there is a constitutive "not knowing where to go," the very definition that he offers of an aporia in the book of that name, *Aporias* (A 12-3). This philosophy of the aporetic wound constitutes both an ethical priority and a transcendental necessity (for the event, if there is such a thing) that problematizes any philosophy of dialectical recuperation or mediation, including pragmatisms and philosophies of the body, both of which he expresses serious misgivings about in all of his work but perhaps especially in *On Touching* as we see in chapter 11.

Secondly, however, the impossible term of the aporia (e.g., absolute, unconditional hospitality) is itself sometimes explicitly described as a wound that haunts the everyday conditional hospitality that we deploy, wherein we provide hospitality to others but only under certain carefully delineated conditions. On this understanding, the impossible prospect of absolute hospitality (or absolute forgiveness) is a wound that haunts our conditional practices of hospitality, and the systems of calculative exchange and political *quid pro quo*. Another important example of this second understanding of the wound occurs in Derrida's recent discussion of the trauma of 9/11 in *Rogues*. Derrida suggests that the trauma that results from an act like the planes targeting the twin towers is not simply an effect that follows from a cause (i.e., the planes hitting the towers and the devastating deathtoll). There is this order of causality, of course, but there is also another order of causality that Derrida might follow the Deleuze of *The Logic of Sense* and call the quasi-cause.[3] Indeed, Derrida goes on to say that the trauma resides not so much in what actually happened (the order of the empirical cause) but more in:

> The undeniable fear or apprehension of a threat that is worse and still to come. The trauma remains traumatising and incurable because it comes from the future. For the *virtual*[4] can also traumatize. Trauma takes place when one is *wounded by a wound that has not yet taken place* ... its temporalisation proceeds from the to come (R 104-5).

It is important to recognize, however, that this logic and this curious temporality are not distinctive to 9/11. For Derrida, all traumas are of this order. In fact, all events are of this order (including friendship as we will see) in which, to emphasize Derrida's point, one is "wounded by a wound that has not yet taken place." We might recall Lewis Carroll's Alice, who cries before pricking herself, and there is a related sense in which we are all wounded by the unknowable prospect of the future. Certainly, for Derrida, any event worthy of the name must involve a confounding of linear time, or the time of the present, including both so-called lived time (produced through habitual syntheses of retention and anticipation), and the time of calculation and prediction (based on the objective time of clocks). There is a future synthesis of time that defies both of these attempts to come to grips with it, and which, while it is wounding, is also the transcendental condition of friendship. Indeed, in precisely the same manner in which Derrida argues that we are wounded by a wound that has not yet taken place, so he suggests in friendship "we are plunged before mourning, into mourning" (PF 14). Let us attempt to come to terms with this transcendental anteriority, this "pre-originary mourning" and what it might involve, via a discussion of his deconstruction of the Aristotelian understanding of friendship.

Friendship and the Mourning of Time

While Derrida's engagement with Aristotle in *Politics of Friendship* eludes pithy summaries, one major feature of it is to insist that Aristotle's famous distinction between perfect (or good) friendships and more utility or pleasure-oriented friendships is untenable due to temporal factors. As Derrida consistently puts it, "it takes time to do without time," which is to suggest that the apparent purity of a friendship (e.g., concern for the other for their own sake, uncontaminated by self-interest) must always have been tested over time (PF 17). We can also ascribe to Derrida a related argument against the stability that a friendship develops over a period of time—although this second argument does not follow quite as well as the first—and to any primacy that is accorded to embodied coping within an environment, in the manner that Heidegger, Merleau-Ponty, and Dreyfus have all differently endorsed, and as we explore in detail in part 3. For Derrida, any friendship must have experienced this futural recognition that whatever arrangements may now be in place (meeting for a beer each Friday afternoon, being open and honest with one another, etc.), they must have passed through an ordeal of time—and one never completely passes through this ordeal—in which these arrangements were subject to revision and to contestation. In other words, to have a reasonably stable friendship one must first have had some kind of experience of the future as unlimited and of nothing dictating that the friendship will continue; this is the trauma of apprenticeship. Moreover, not only does a stable friendship need to traverse this unstable order of time in which this openness toward the future is experienced unabated, but Derrida also intimates that even after having endured this trial of friendship, we still *ought* to think (and to live) friendship with an "open heart," open to the "perhaps" which engages the only possible thought of the event and the future (PF 16, 29-30). The aporetic and traumatic is hence not only the condition of friendship, normality, and the like, but it also serves as an ethico-political injunction that might be formulated as follows: respect the aporetic dimension of friendship; do not cover over the wound. Indeed, Derrida repeatedly describes recognizing, rather than covering over the aporia, as a condition of responsibility (cf. A 16).

We will return to this, but contrary to the famous Aristotelian account of friendship in book 8 of the *Nichomachean Ethics*, Derrida hence argues that friendship *should not* be too stable, and should not railroad the future into its habitual expectations, which is to deny not only the difference of the other person, but also the more radical difference that time insinuates into any and every friendship. According to his quasi-transcendental argument (it is quasi-transcendental because it also renders pure friendship of the Aristotelian variety impossible), the apparently stable aspect of what we might call "lived friendship" is always broken open by this radical future, this temporal wound (which we mourn) that paradoxically precedes friendship, firstly in the trial and the test, in the selecting and preferring, in the question and the objection. Without this, there is no friendship but mere robotic acquiescence, and yet, at the same time, this test also interrupts any neat and easy distinction between pure (perfect) friendship and its more utility oriented manifestations. According to Derrida, it even problematizes any too-easy friend/enemy distinction, upon which the political philosophy of Carl Schmitt relies, as has the foreign policies of successive United States governments in their consistent refrains concerning rogue states and George W. Bush's now famous "axis of evil." Indeed, Derrida suggests that: "The possibility, the meaning and the phenomenon of friendship would never appear unless the figure of the enemy had already called them up in advance, had indeed put to them the question or objection of the friend, a wounding question, a question of wound. No friend without the possible wound" (PF 153).

On what basis might Derrida begin to justify such a claim? Well, if Aristotle is right that we cannot have too many friends (and Derrida thinks he is, because we cannot spend the required amount of time and attention with each), this complicates Aristotle's own more general claim that genuine friendship involves loving the other person for their own sake, and not just for the utility or pleasure that we might derive from the friendship. After all, friendship takes time to develop, and this time cannot be given to everyone. We cannot be genuine friends, in the Aristotelian sense, with lots of people. For Derrida, there is hence a type of oligarchy at the heart of friendship and it is not the great model for democracy that Aristotle and others have suggested it is. This is so even without considering those who were overtly excluded from the Athenian fraternal band of friends—women and slaves—because Derrida also shows the subtle privileging of fraternity, brotherhood, and ties of allegiance and filiation that have surrounded the Western history of ruminations on friendship. But to return to Derrida's key point, despite this oligarchy at the heart of friendship we nonetheless generally want the most and best forms of friendship possible, and we hence must prefer certain friends to others, and make choices between who will, and who will not be, our friends. Among other things, this means that we must calculate with our friends, which is precisely what Aristotle argues that we ought not do with good (or perfect) friends. For Aristotle, we do not, and should not, use them as means to our own ends and try to deduce their worth to us, but, for Derrida, this is inevitable because we must prefer certain friends to others. We must put the chosen one(s) through the test by living with them, and in order to establish who we will try this with, there is always a choice, a decision, and it involves (as well as an undecidable leap) calculating whether or not these prospective friends are good for us personally and give us pleasure. Our focus cannot be restricted to whether or not a friend is good in him or herself, as Aristotle's position on perfect friendship suggests, because even if we imagine ourselves to be a good citizen seeking *eudaimonia* and the cultivation of our virtues, there are nonetheless likely to be plenty of people who we recognize as being good but whom we do not want to be friends with. For Derrida then, Aristotle's distinction is too quick: friendship requires that the friend be good in themselves (and hence loved for their own sake), and also for them to be good for us (and hence give us pleasure and utility), which requires prudential judgment about what is in our interests. Thus far I am entirely in agreement with Derrida: time does problematize any idealistic account of perfect friendship like Aristotle's.[5] Whether it equally successfully problematizes the virtue that Aristotle associates with a stable friendship (i.e., what Derrida refers to as the "omni-temporal" relationship that we acquire with the world and others through a process of *hexis* and *phronesis*), however, is not so clear, and even less clear is whether the establishing of any such transcendental priority also legitimates the ethical impetus that Derrida also gives it in his explicit and implicit suggestions that true friendship ought to be both open to, and oriented around (recall his phrase an "open heart") to the temporal wound that is its condition.

Less commonly acknowledged but perhaps even more important for my argument that Derrida's analysis of friendship is "touched by time" in sometimes problematic ways, is the manner in which *Politics of Friendship* also anticipates certain themes that came to prominence in other books like *Adieu to Emmanuel Levinas, Work of Mourning*, and that were already apparent in *Memoires for Paul de Man*.[6] In short, for Derrida, friendship is inexorably bound up with mourning. While it might be protested that mourning is merely a structural possibility of friendship rather than a necessity as the above quotation from Derrida seems to suggest—"no friend without the *possible* wound"—it is important to recognize that this structural possibility contaminates, and is the condition for, anything that happens. As Derrida explicitly suggests:

> Here again, the difference between the effective and the virtual, between mourning and its possibility, seems fragile and porous... The anguished apprehension of mourning (without which the act of friendship would not spring forth in its very energy) insinuates itself a priori and anticipates itself; it haunts and plunges the friend, before mourning, into mourning (PF 14).

As we saw with his account of the trauma of 9/11 ("trauma takes place when one is wounded by a wound that has not yet taken place"), the event of friendship only takes place when (and if) its temporalization proceeds from, and is wounded by, the future that is still "to come." This is a necessary but not sufficient condition for friendship, in that without this wounding or haunting we will merely be indifferent toward the other person, never passing beyond a relationship of casual acquaintance (there is hence a minimal sense in which Derrida agrees with Aristotle, even while he points to the way Aristotle's argument simultaneously undermines itself).

We can also understand this originary mourning in the Levinasian sense that one mourns the death of the other (and is responsible to them) in advance of their actual death. Indeed, like Levinas, Derrida insists that the more fundamental wound is the prospect of the other's death, rather than one's own death, as in Heidegger's famous account of *Dasein* resolutely confronting its "being-towards-death" in *Being and Time*: the death of the other is the first death, ontologically or metaphysically speaking. But Derrida also means more than this, and seeks to leave this dispute between Levinas and Heidegger behind by complicating the binary self-other logic of their respective positions on death. Indeed, he comments in *Aporias* that neither Heidegger, Freud, nor Levinas, take into account the originary mourning that is his primary concern (A 39). Derrida explicitly states that if the feeling of mineness (*Jemeinigkeit*) or self-identity is constituted by an originary mourning, as he thinks it is, then "the relation to the other (in itself outside myself, outside myself in myself) will never be distinguishable from a bereaved apprehension: The relevance of the question of knowing whether it is from one's own proper death or from the other's death that the relation to death is instituted is thus limited from the start" (A 61). The argument here is clearly an expanded one: it is not only friendship, but any relation to the other involves some kind of bereaved apprehension or originary mourning. The warrant for this extension is not immediately apparent, but this is a curiously melancholic sentiment. If there is also an ethical impetus bound up with this wounding aspect of time (the future that haunts), as I have suggested that there is, then it is in this sense that Derrida's work can be said to institute a transcendental pathology; one aspect of time—its wounding component which we mourn—has both a transcendental priority and an ethical privilege over the time that binds, the omni-temporal time that scars. In fact, this is my key claim in this chapter so it is worth considering in greater detail how this transcendental/ethical obfuscation might take place in Derrida's work.

These subtly different understandings of "originary mourning"[7] as peculiar to the event of friendship, or as a condition of any relation to the other at all, correspond to a problem upon which Derrida's philosophy is avowedly situated, but which nonetheless consistently involves the obfuscation of the transcendental and the normative, and thus, on my view, ultimately makes possible Derrida's ethics of nonpresentist time. For Derrida, any relation to the other must involve a pre-originary mourning, and yet he also argues that friendship, if there is such a thing, is premised on this pre-originary mourning and ought to embrace rather than turn away from this condition: the former claim is Derrida weighing into more traditional transcendental philosophy; the latter claim also more clearly gives this mourning an ethical and normative dimension.

This dual status of his key transcendental claims recurs throughout many of his major concepts. There is an unresolved tension in Derrida's work, for example, between whether events are to be understood to necessarily happen all of the time (in the discussions of iterability and difference in his early work this seems to be the case), and yet he is also prone to employing a more genealogical logic which suggests that such and such must be the case for any "event worthy of the name," often with an added clause such as, "if there is such a thing as an event." Interestingly, Derrida also seems to suggest that there is, in fact, a necessary overcoming or confusion of the is/ought divide in any event worthy of the name. This is perhaps clearest in his discussions of the United States declaration of independence in "Otobiographies," along with his essay on this same theme in *Negotiations.*[8] The declaration of independence seeks to describe or ratify an already existing state of affairs, but at the same time its validity is invented through this act of naming, this performance, which also prescribes what ought to be the case. Consider the famous passage:

> We therefore the representatives of the United States of America, in General Congress assembled, appealing to the Supreme Judge of the world for the rectitude of our intentions, do, in the Name, and by the authority of the good people of these Colonies, solemnly declare and publish that these United Colonies are and of right ought to be free and independent states.

Manifest in this statement is a description of a state of affairs (the assembled people, even though they are not yet legitimately representatives), and the imperative regarding what ought to be the case (their being representatives of free and independent states), and it is the performative aspect of the declaration that makes possible the transition from the one to the other in a relatively seamless way.[9] This analysis of Derrida's is insightful, and it bears some relationship to his account of friendship as we will see, but the movement between the "is" and the "ought" that features here also persists in more troubling ways in Derrida's own work.

A similar structure undergirds his work on the decision, another example of a concept that is simultaneously impossible within its own internal logic and yet nevertheless necessary. He sometimes suggests things like "the decision, if there is such a thing as a decision"[10] and hence can claim that he is merely analyzing the logic of the decision. More commonly, he asserts that the various decisions that we do make must be structured by the experience of the undecidable (which is thus the transcendental condition of any particular decisions that we do make). Quite frequently, however, he also argues, or implies, that our actual concrete decisions not only ought to be structured by this experience but they ought to embrace, rather than turn away from, this condition.[11] This formulation allows for degrees, and an ethico-political orientation that the transcendental argument itself should not admit. In "Ethics and Politics Today," Derrida makes this explicit as we saw in chapter 4. He comments that the strategy of deconstruction, and its deployment of transcendental arguments should be considered to be primarily "pre-ethical-political," but even there he cannot resist going on to add that it in fact simultaneously *is* ethico-political, precisely because it prescriptively insists on this preliminary "pre" that *must* accompany all responsible decision-making. Here is where the normative comes in, a normativity that paradoxically grants an ethico-political privilege to a transcendental condition over what we might call the extra-transcendental, or the empirical. The problem, then, revolves around whether we understand Derrida primarily as a genealogist of concepts, as an undisguised transcendental philosopher, or as an ethicist, albeit in a highly restricted sense. For me, however, he trades on this ambiguity and it allows him to imbue this transcendental temporal wound with a normative dimension that is never satisfactorily justified. As such, I think Derrida makes almost precisely the same mistake he accuses Heidegger of when in "*Ousia* and *Grammē*" he rejects Heidegger's quasi-ethical distinction between authentic and vulgar time.

I will continue to try and bring this out in regard to the relationship between time and friendship. What, after all, does Derrida's paradoxical suggestion that without a certain kind of transcendental mourning (the wound) there is no friendship mean? How might one be plunged before mourning into mourning? As well as the prospect of the death of the other, it seems that at a more fundamental level what we are mourning is time, or to phrase this another and perhaps better way, the mourning that traverses us is a certain kind of undecidable experience/non experience of time. Although this cannot be understood to entail a mere lament that time is passing (which would be *ressentiment*), without (typically) being explicitly aware of it we necessarily mourn the future, since it is from the future that death, trauma, loss, come. Derrida's position seems to be that what we mourn, in pre-originary mourning, is that we and our loved ones will never arrive there (at death) together, *rendezvous*. We are waiting for each other in this contretemps of mourning, and there is a sense in which, for Derrida, this waiting for something that is impossible is the condition for mineness and self-identity. Moreover, we have already seen how time disrupts and problematizes friendship. The time of stable and habitual friendship is the time of confidence and chronology; Derrida's etymology of the term confidence draws outs its links with fidelity, faithfulness, the sensible duration of time. But it takes time to do without time in this manner, it takes trials and tests, and these memorial wounds are always there and never dispensed with (PF 14-5). Indeed, Derrida emphasizes that a condition of stable friendship, of what he calls "cultivated aptitude," is this *contretemps* that is associated with both the past and the future (PF 16). This condition of (im)possibility of friendship is the time of the irruptive wound, which "gives itself only its withdrawal," and disjoins presentist time while it also pertains to it (PF 14).

In this context, it is also worth reflecting on that famous statement attributed to Aristotle by many since the middle ages, but always without any concrete referent, and from which Derrida begins each of his chapters in *Politics of Friendship*: "O my friends, there is no friend." This paradoxical statement seems to undermine the tenor of Aristotle's work on friendship in the *Nichomachean Ethics*, and it is also important to Derrida because it is split between two temporal orders and hence not a simple contradiction: the first clause invokes those friends who have been or those friends who will be (its performative element either calls them forth or recalls them—this is the time of the future and the past); the second clause is irremediably of the present in its assertion of the definitive, factual statement that there *is* no friend. There is no good or perfect friend, because friendship is split between these orders, never unambiguously present and self-contained. It is haunted by the trial with which it developed and began (memorial time), as well by the future (the death of the other). We mourn the split or wound of time, that is to say the way in which friendship is never finally given but must be re-performed and re-invigorated, and also the way in which friendship is never pure but always also compromised and liable to transformation, including the possibility of an unforgivable rupture.

At the same time, in a certain sense we must also celebrate this wound, however, for without it no friendship is possible. Why Derrida feels compelled to use loaded terms like "mourning" and "wound" for this aspect of time hence remains an open question, albeit one that is more explicable if one agrees with my argument that Derrida's philosophy of time instantiates a transcendental pathology, in that a surreptitious ethics (with a melancholic inflection) is derived from a transcendental analysis. On his view, however, this split is also what makes possible the event, the new, and difference, to momentarily treat them as synonyms. As Derrida suggests, "the disposition, the aptitude, even the wish—everything that makes friendship possible and prepares it—does not suffice for friendship, for friendship in act ... the analysis of conditions of possibility, even existential ones (e.g., habit), will never suffice in giving an account of the act or the event" (PF 17). Indeed, Derrida's basic critique of habit, skills and *phronesis* (as they apply to friendship) in *Politics of Friendship* is, as we have seen, that they are "omni-temporal" (PF 16, 20), and cover over the gap between two different temporal orders.

And it seems to me that Derrida is right in this respect; corporeal responses to our friends do form habitual patterns, and the quasi-transcendental time that haunts friendship (the wound that come from both the past and the prospect of an unknown future) is covered over in our typical phenomenological experiences of friendship.[12] However, it should be noted that these patterns that any friendship partakes in, far from suggesting any kind of mechanistic stagnation, are in fact what allow us to intuitively and prereflectively notice that something is different or amiss with a friend, and thus to transform our relation to the individual in question. This applies to moral considerations pertaining to friendship in a structurally isomorphic way—we are also ahead of ourselves, awaiting our friends in *Mitsein*, in advance of any trial—to our general coping with equipment and objects in the world, even though the *tēlos* of our relations with others is structured rather more by desire for the unknown than imminent equilibria and ease of use. Aristotle, Merleau-Ponty, Dreyfus, and many others have shown us that these corporeal adjustments toward others involve a sophisticated feedback loop and are not merely mechanistic and unthinking reactions that pay no attention to the distinctiveness or alterity of the other; in fact, they enable the other's particularity (and their needs) to come forth. Practical wisdom just is this ability, which can be developed over a period of time, to respond in the appropriate manner to the consistently changing circumstances that one is presented with; it is based on the development of skills, moral virtues, and the like within a community. This kind of embodied and prereflective responsiveness is itself a quasi-transcendental condition of friendship of sorts. And while it is true that such capacities take time to develop (that is, it takes time for us to become morally mature and suitably attentive to the specificities of our friend and their situation), it is also the case that they are not something that we acquire *ex nihilo*; such abilities are always already at work in any relation to the other, and in friendship. My point in this regard is a simple one: although this omni-temporal time "does not suffice for friendship," as Derrida compellingly suggests, it is nonetheless as central and important to friendship as is the split or wound that he more forcefully insists upon.

More critically for my purposes here, it also remains unclear and something akin to an unjustified point of faith in Derrida's work as to why the temporal wound in friendship should have a normative and ethico-political impetus (respect the wound, be open to the "perhaps," etc.) that is missing from his occasional descriptions of more existential and embodied conditions of possibility. One of Derrida's interviews published in *A Taste for the Secret* is revelatory in this regard. As well as affirming the importance of Kierkegaard to his own work, he suggests that his intention was never "to draw away from the concern for existence itself, for concrete personal commitment, or from the existential pathos that, in a sense, I have never lost ... In some ways, *a philosopher without the ethico-existential pathos does not interest me very much.*"[13] In the case of the major existentialist thinkers, their phenomeno-logical and existential descriptions gave this "ethico-existential pathos" some content, and in the more subtle and nuanced of these thinkers they were al-ways moderated by a recognition of the importance of bodily equilibria, the ready-to-hand, etc. In taking the transcendental turn in the manner that he does, Derrida's "existentialism" is one without either of these anchors.[14] As such, his is an ethical *pathos* enshrined by a transcendental philosophy (and, tacitly, a hierarchy) of time. Or, if we think he remains faithful to his early declaration that there is no concept of time other than the metaphysical one, it is a hierarchy of the untimely on the one hand, and the metaphysics of pres-ence that it disrupts on the other hand. Of course, that is not to deny that, for Derrida, the two hands must be intertwined, in that the former *contretemps* must always also pertain to time, especially in the manner that the future and past haunt and interrupt any metaphysics of the present. Nonetheless, that does not mitigate the lack of attention that Derrida accords to corporeal phronesis, which is arguably not reducible to a metaphysics of presence in any case,[15] and it does not account for the manner in which his brief refer-ences to it always conspicuously lack the ethical flavor that accompanies his descriptions of the time of the past and the future, which are, for him, the transcendental conditions for the present and for friendship.

A certain wound of time that Derrida has given various names operates as both a transcendental condition and, at times, as a hierarchical and privileged ethical term. Despite his forceful analyses of the contamination that time in-stitutes, it is far from clear that this justifies an ethical priority to be accorded to but one aspect of time, the disjunctive and the wounding. Derrida's argu-ments toward this conclusion too often beg the question and assume the im-portance of this temporal disruption. But what kind of ethics and politics does this valorization of the event as rupture (as outside of the order of possibility) have? Is it sufficient? Clearly not, even according to Derrida for any concrete political action as we saw in chapter 4, and it is not sufficient for an account of friendship. Indeed, in both of these respects phenomenologies of the body are central: the fact that the body excludes things from our particular horizons of significance is not something that should be ignored, nor should the way in which they involve an omni-temporal binding that covers over gaps and aporias, without suggesting they are not there. Time, it seems, unsurprisingly enough, is complicated. And while it would be difficult to deny that Derrida grasps this complication, I hope to have shown that he nonetheless imports an a priori judgment about which aspects of time are most significant and of the most value. While time wounds us in the senses enumerated so well by Derrida (and Deleuze), lived time also scarifies these wounds, covers them over through an omni-temporal time that conjoins, and this is not unimportant to friendship. In fact, we might even reverse Derrida's formulations and say that there is no friendship without scarification, without a confused and ambiguous present, something that is occluded and downplayed in Derrida's temporal and ethical decisions.

Part Three

Phenomenology, Embodiment, and Pragmatic Temporality: An Anachronistic Dialogue

9

Time Out of Joint: Between Phenomenology and Poststructuralism[1]

In this chapter I seek to highlight in a preliminary manner one of the core major differences within continental philosophy in regard to the philosophy of time—differences that I associate with, respectively, phenomenology and poststructuralism (including the renewed interest in, and use of, Nietzsche and Bergson's work by poststructuralist philosophers). This tête-à-tête concerns the connection between such philosophies of time and metaphilosophical and ethico-political normativity. Indeed, it is no coincidence that in some of the canonical rejections of phenomenology, such as those proffered in Meillassoux's *After Finitude*, Deleuze's *Difference and Repetition*, *The Logic of Sense*, and *What Is Philosophy?* (with Guattari), and in Derrida's *Speech and Phenomena* and, more recently, *On Touching: Jean-Luc Nancy*, the most telling and repeatedly expressed objections about are about time and transcendental philosophy. In *Speech and Phenomena*, for example, Derrida shows that the emphasis upon the centrality of the living-present in Husserl's work is intimately related to what Husserl calls phenomenology's "principle of all principles: that every originary presentive intuition is a legitimizing source of cognition."[2] As Husserl goes on to say, everything that presents itself to consciousness is taken in the manner that it presents itself, and Derrida has questioned phenomenology for its tacit dependence on a "now" moment, and a temporal immediacy of that which presents itself, remarking that "in the last analysis, what is at stake is . . . the privilege of the actual present, the now."[3] But to put Derrida's concerns in a more general context, the worry is that phenomenological descriptions of the experience of time focus, predominantly if not exclusively, on the manner in which time gathers, or conjoins rather than disjoins (and we will see that this appears to be true of authentic *Dasein*). "Lived time" is described by phenomenologists (like Husserl and Merleau-Ponty) as a neat and unified continuum, but for the poststructuralists this kind of experience is an illusion of sorts. On their view, the unity of experience revealed in the living-present covers over something more fundamental about time—if I can put the point somewhat dramatically as the theorists involved typically do, that is time as wounding (see Deleuze's *The Logic of Sense*), time as out of joint in the manner of Hamlet's memorable refrain (see Deleuze's *Difference and Repetition* and Derrida's *Spectres of Marx*), time as "nick" as Elizabeth Grosz puts it in *The Nick of Time*, or time as ungrounding and casting asunder the identity of subjects and bodies.[4] Phenomenological accounts of time are also thought to be problematic for still seeing time as, if not the measure of change, then at least as irremediably bound up with movement. In opposition to this, the poststructuralist philosophers want to quite radically *disassociate* time from movement. What is at stake in this charge, and why would it matter if it is true? The poststructuralists allege that any association of time and movement threatens to be unable to explain the advent of genuine difference and novelty, and is, at best, only an indirect way of understanding time.

Of course, one would not want to overstate the differences between phenomenologists and poststructuralists. We might note that in certain writings Derrida is not clearly on what I am characterizing as the poststructuralist side of the equation at all. In addition, poststructuralist understandings of time crucially depend upon the work of phenomenologists like Heidegger and Levinas as much as they depend on Nietzsche and Bergson, and they all share in common the concern to avoiding reducing philosophy of time to clarifications regarding the physicist's understanding of it (Einstein and four-dimensionalism).[5] Moreover, both phenomenologists and poststructuralists also seek to avoid a conception of time that we might associate with common sense and the natural attitude. This view understands time as the chronological succession of an infinite series of instants, stretching from the future to the past. This conception of time involves a series of moments, and a linear trajectory, which clock-time regulates and subjects to measurement. The arguments against such views tend to rely on various forms of transcendental reasoning, which attempt to show either that linear-clock time—or theoretical ideas of time that are dependent on a clock-like series of moments—is an abstraction from lived time, or that clock-time presupposes the existence of a past that cannot be recalled and a future that cannot be anticipated, and thus give us a one-sided account of the structures of time.

But to return to the *differend* that separates much work in contemporary continental philosophy (as practiced both on the continent and in Anglo-American countries), while I think that Derrida, Deleuze, and others are roughly correct in their diagnosis of phenomenology's association of time (or, better, temporality) with the subject, including with the movement of subject, I am not convinced that they are correct in considering this to be a theoretical weakness, nor that the proffered alternatives are to be preferred. In a critical vein here, I point to some problems with the "time out of joint" perspective of Deleuze, Derrida, and others that will be developed in subsequent chapters. My basic worry is that too often the transcendental critique of any emphasis on the "living-present" and other such "chronopathologies" trades on claims of necessity that are either speculative (the transcendental claim is not established as a necessary one, but is at best a weak inference to a better explanation) or that depend upon their association with an accompanying moral and political tenor (what I will call "empirico-romanticism" in the next chapter) that threatens to be dogmatic. While I agree with them that time and politics are intimately connected, I also think that theoretical accounts of this fragile connection need to be careful to avoid lapsing into dogmatism, and this is so even if the relevant conception of time is not tethered to any teleological account of the trajectory of history. To worry about this risk is not to simply be the victim of a false problem as Nathan Widder suggests, or a transcendental "illusion" as James Williams phrased a related objection.[6] But this is all very abstract. After a brief exposition of Heidegger's account of pragmatic temporality in *Being and Time*, a text which is central to both of the key continental trajectories that I am exploring in this book, I will then use a simple sporting example to try to clarify some of the key aspects of a phenomenological account of temporal experience, as well as what Deleuze and others are worried about in phenomenology's focus on both lived-time and the living-present.

Heidegger: Vulgar Time, Pragmatic Time, and Originary Temporality

Being and Time has a remarkably complicated account of the structures of time that I will not be able to do justice to here. That said, it is arguable that the various orders of presupposition are so complex that one should perhaps pause before accepting any such intricate chain of layers in a transcendental argument, since each condition is only as plausible as the others with which it is associated.[7] Nonetheless, it is necessary to provide some kind of indication of the differences in *Being and Time* between respectively: originary temporality (with its three temporal ecstases and the priority of the future), world-time (which is itself conditioned by pragmatic time, the latter of which partakes in both originary temporality—in the originary present—and world-time), and ordinary or vulgar time (as well as scientific elaborations on it). The terms themselves probably immediately disclose the orders of priority that Heidegger argues obtain between each of these aspects of temporality, and his basic idea is that one can best explain the most important features of ordinary time by appealing to worldly time (in particular, through pragmatic time and its breakdown when we need to "reckon" with time), which can in turn be explained by appealing to originary temporality. Part of the point behind such reasoning is to acknowledge that the ordinary conception of time (as well as philosophical and scientific elaborations on it) has some phenomenological basis, but that it is limited and circumscribed in particular ways. Of course, this chain of transcendental reasoning is not entirely neutral in regard to the distinction between authentic and inauthentic modes of existence with which Heidegger's entire book is concerned, and which remains a focal point of poststructuralist critiques of it. Certainly there are several passages in *Being and Time* that clearly associate so-called vulgar time and inauthenticity, and sometimes the descriptions of original temporality have a connection with authenticity, as is clearest in the discussions of being-towards-death that we consider in chapter 12.

But what is vulgar or ordinary time? Vulgar time supposes that time is essentially measurable by clocks, and that it is something that is independent of all contexts (and moods). Heidgger maintains that we do have a phenomenological experience of time as spanless instants without content that are independent of us, but this ordinary time is time conceived as *Vorhandenheit*, as present-at-hand, and it is said to be inauthentic because it lacks temporal unity. It gives us a disengaged or purely formal "now," without what is typically translated as "significance" and "datability." Scientific time, for Heidegger, is just ordinary time supplemented by mathematics and physics. As such, it is not false or illusory but partial, although it must be recognized that sometimes Heidegger does come close to suggesting that this kind of relation to time is motivated by the desire to flee from the finitude of temporality as revealed by the experience of anxiety (*Angst*), in which all of our habitual and everyday ways of relating to the world (and worldly time) drop away and sink into insignificance. Forced to confront our own thrownness and finitude, anxiety individualizes us because we no longer feel at home in the world of the ready-to-hand, of *das Man*, and the many.

Now, Heidegger claims that ordinary time is a derived version of what he calls world-time, the latter of which is said to be modally indifferent between the authentic or inauthentic. World-time involves at least four key features: "datability," "spannedness," "publicness," and "significance" (BT §79). By datability, Heidegger means that world-time associates the past with particular things that happened then (e.g., formerly, when I was a student reading Heidegger for the first time). This is a past-present (a no-longer-now), and world-time likewise posits only a future-present (a not-yet-now). The more general point would be that, as William Blattner puts it, "in its everyday dealings in the world, *Dasein* relates to times as contentful, as 'times, when x.'"[8] And on such a view, there is a dominance of the "now." This experience of time can also "level off" into ordinary time, but we typically do not experience the passing of seconds or the passage of time in a disinterested and disengaged manner; moreover, when we do, Heidegger says, it is experienced as an interruption to worldly time and presupposes that background. In worldly time, the now is not an instant, but is spanned in the sense that it stretches out depending on a given pragmatic context, and we also experience time as having significance and normativity, both when we are involving in reckoning with world-time and in what Heidegger calls pragmatic time. In regard to the latter, a given "now" shows up as that time in which we ought to hurry to get to the lecture theater before the class begins (and there is no need for reasoning about this), or as demanding a certain kind of pass of the basketball to get it to a teammate without it being intercepted by the opposing players. In this kind of pragmatic now, we also have an orientation to the future as well as the past. We are ahead of ourselves at the completion of the task, awaiting and anticipating its completion, while we also retain and presuppose a background that includes the relevant bits of equipment for use in any given task, one's bodily orientation to them, as well as procedural memory about how to accomplish certain tasks. Heidegger suggests that such pragmatic temporality is thus a condition for our apprehension of worldly time (BT §79). But if worldly time is sequential as Heidegger suggests it is (indeed, this is the feature that primarily distinguishes it from original time), and involves the positing of a series of "nows," how do we get from the dominance of the "pragmatic now" to the sequentiality that is characteristic of worldly time proper?

Heidegger's solution is to suggest that in situations of breakdown and when things are not working in the manner that we are accustomed to them working, we must interpret or "reckon" with time, rather than being precognitively engaged with tasks and equipment that have pragmatic time as their (immediate) condition. In this manner, we become aware of time as a whole sequence of nows, are able to plan courses of action, calculate how best to secure certain goals, and so on. This is the difference between being immersed in an activity, and then having to step out of this immersion to analyse a given activity and its temporal ordering (e.g., will my lasagna be ready for the guests who are coming for dinner this evening; how do I make sure this is so? Time is passing and I am not getting anywhere!). William Blattner nicely summarizes the relation between worldly time and pragmatic time in the following terms: "we are concerned with time, and the temporality of concern is pragmatic temporality"; and later, "reckoning with time is reckoning with Nows, and pragmatic temporality is the fundamental capacity through which the framework of the Now is available to *Dasein*."[9]

Now, thus far I am generally in agreement with Heidegger. *Dasein* is able to understand world-time only because it is a temporal entity. And although Heidegger does not pursue this direction of thought, it is arguably our proprioceptive sense of our bodies (and hence time cannot be privileged over space, contra Heidegger), and the habits and skills that are thus made possible that is fundamental to this pragmatic temporality. As Shaun Gallagher argues, there is a sense of mineness (*Jemeinigkeit*) involved in proprioception that is the condition for both a proto-self-consciousness and proto-time-consciousness, and from which self-reflection and other adult capacities develop but never entirely leave behind.[10] But, of course, Heidegger does not stop here. For him, pragmatic temporality arises out of originary temporality, in the sense that originary temporality explains the various features of world-time, including its pragmatic dimension. As Blattner says, "the originary future (the not-yet of abilities) explains the pragmatic future (the not-yet of tasks) and the originary past (the alreadiness that characterizes the way things matter to *Dasein*) explains the pragmatic past (the alreadiness of the wherewithal's making a difference)."[11] In other words, Heidegger's point in relation to the pragmatic future is that we aim at fulfilling certain tasks because we have something we want to be or become. In the Heideggerian parlance, there is a "for-the-sake-of-which" that explains particular pragmatic ends. So, pragmatic futurity requires originary futurity and, Heidegger contends, the reverse is not the case: hence the priority of originary time. This claim seems to depend on the following sort of reasoning: if I were to ask myself why I am producing this book, it is natural to say that it is because I understand myself as having something interesting philosophically to say and I care about the prospect of being a decent philosopher (in Heidegger's sense of care, i.e., concern). But is it also natural to say that I have a self-understanding that I am a philosopher because I am engaged in the task of writing this book? Heidegger says no, but I think that this is not so clear. In this respect I am perhaps more on the side of Sartre, who in *Being and Nothingness* argues that situation and motivation for pursuing certain projects are indistinguishable.[12] Perhaps Sartre's claim is overly dramatic, since we conceptually can and do distinguish between situation and motivation (and between pragmatic time and originary time), but my claim is that in concrete action there is a mutual dependence between pragmatic time and originary time. I do care about philosophy because I have read and written a lot, and this has led me to see philosophy as of interest and to appreciate (to some extent!) the subtleties of the discipline. So, unlike Heidegger, I see reciprocal dependence here rather than a neat transcendental chain of presupposition. Moreover, genetically speaking, we might also ask, what comes first, proprioception and bodily motility, or the "for-the-sake-of-which," the "not-yet" of abilities? To some extent the genetic question is not Heidegger's concern in his transcendental philosophy, but the claim that *Dasein* can discover entities and equipment in relation to tasks (and hence the institution of worldly time) only because it has aims "for-the-sake-of-which" seems empirically wrong, and, on its own terms, it also raises some complicated problems regarding the relation between animal and human life more generally that we will return to in chapter 12. Indeed, in conjunction with his analyses of death and their later temporal recapitulation, Heidegger maintains that for authentic *Dasein* time passes in a coherent and connected manner—that is, at least once *Angst* has jolted *Dasein* from its immersion in ordinary time, which is an inauthentic derivation from worldly time. He also argues that authentic temporality is "resolute" because we have a life-project that gives us meaning—and temporal unity—which is premised on the recognition that we are not immortal (BT §§81, 61). These aspects of the Heideggerian account of time are ones that we will have cause to problematize, but I have broadly endorsed his pragmatic account of time and will develop that in the remainder of this chapter, while also attempting to show its compatibility with other phenomenological accounts of the living-present, particularly Husserl and Merleau-Ponty's.

Cricket and the Living-Present

It is received wisdom in cricket and other sports that players both are not, and should not, be directly phenomenologically aware of any kind of conscious decision-making processes while absorbed in what various theorists since Hubert Dreyfus have called skilful coping. In cricket, one reason for this kind of injunction is obvious enough: batting is, as John Sutton observes, regulated improvisation under severe time constraints.[13] Faced with a fast bowler (for those from the United States, think of an express baseball pitcher), say Brett Lee in his prime, there is no time for thinking or any kind of hesitation; batsmen need to spontaneously respond and to be totally absorbed in the moment. There is not even time, according to Sutton's research, to actually watch the ball all the way and then respond. Despite the fact that almost all cricket coaches will advocate unwavering watching of the ball, elite players do not watch the ball for its entire trajectory. The best players watch it out of the hand, anticipate where it will land and direct their vision there, then attend to where it lands on the pitch and anticipate where it will go. Without this kind of anticipation one could never response adequately to the visual stimulus in a timely fashion when faced with a 150 km per hour delivery.

Consider the temporal experience involved here. It seems clear that in any living-present the sports player retains the past in the form of a retention or sedimentation in the body of what has happened before. At the same time, they must also anticipate probable future scenarios regarding what will be likely to happen in the future. Such coping techniques simultaneously carry the weight of past sedimentation and yet are also productive of a world of anticipated possibilities that are increasingly differentiated from each other; for the expert, the situation solicits increasingly refined responses. Moreover it is the ability to perform such anticipations more quickly, and with greater accuracy, that separates experts from those who are merely competent. Such responses cannot be mechanistic, or rigidly rule-governed. Every stroke will need to be played in slightly different circumstances, on a different pitch, with differing wind conditions, differing condition of the ball, and an altered trajectory of the delivery, to mention just a few of the variables. As such, any given cricket stroke will never be totally new, but neither will it be brute or instinctual repetition either, having to be attentive to the difference presented by each ball, but still implicitly drawing on one's repertoire of past experiences that contribute to each shot (hence each batsman has a recognizable style). Through training and skill, one is solicited by the situation to respond to it in more and more nuanced and specific ways. Being-in-the-present, on this view, involves an experience of time that synthesizes or integrates elements of the past and the future within its purview.

In his reflections on the phenomenology of internal time-consciousness in the book of that name, Husserl makes a related point. He famously suggests that our integrated experience of a melody—even on first listening—implies that any so-called "now" must have a retentive element that retains the past notes, and a protentive moment that anticipates future elaborations, as well as what he calls the primal impression. Otherwise, our experience of the sounds would be random and disparate in a way that it is not, without any kind of ability to hear a melody.[14] This kind of temporal experience then, is a synthesis, in which any living-present, for it to be meaningful, involves a retentive and protentive element rather than being a self-contained instant, or a series of such instants. While Heidegger criticizes Husserl's account of time as a disengaged awaiting, and thus involving a disengaged "now" (stripped of context), this seems unfair. Whatever we think about Husserl's emphasis upon phenomenological *reflection* and its distance from Heidegger's more pragmatic (and existential) phenomenology, his account of time-consciousness is not radically different from the account of pragmatic temporality that Heidegger himself offers, and it is amenable to further socio-historical contextualization as Husserl came to emphasize in his later work.

While there is an important phenomenological distinction between these kind of acts that involve procedural memory and a passive synthesis of time, and explicit biographical memory or reflection on our past, we should also note that the "don't over-think" injunction that plays a large role in sporting activity usually prescribes more than just being in the present when engaged in the activity in question. The cricketer who dwells on the past between deliveries, or who reflects on their lucky escape a ball before when someone dropped them on 99, or the minefield that is the deteriorating condition of the pitch, is not likely to perform well. Likewise the player who is preoccupied with getting to the lunch break in ten minutes without being dismissed, rather than playing each ball on its merits, is also likely to make a mistake.

This does not seem merely to be folklore, but is borne out by various studies.[15] An elite sportsman or woman thus has to train themselves to put past biographical experiences out of their mind; certainly out of their immediate focus. Being a good cricketer depends not merely on talent and training in the various skill domains, but on training one's mind; in particular, in controlling one's temporal experience throughout an afternoon or so of projectiles being aimed at one's torso. Without putting too fine a point on it, it seems that one performs better when one is not haunted by ghosts, which we might follow Derrida and note always come from the past and the future.

Likewise, Dreyfus produces some quite compelling empirical research on decision-making processes that suggest that it is a spontaneous embodied responsive to the environment (which is not a matter of rational calculation) that leads to mastery and expertise in any number of given fields, whether they be basketball, chess, business, or even morality.[16] We look at this in more detail in the next chapter, but for the moment it suffices to say that constant calculators, and people who reflect all of the time on the best course of action to take, tend not to make the best decisions and do not often reach the highest levels of expertise in a given field. This suggests that expert activity involves a disciplining of the manner in which one experiences time. There is, we might say, a kind of expert-induced amnesia, which is actually not necessarily a weakness. As such, we have an account of the synthesis of the living-present, as well as a normative account of how to "successfully" live time, at least in relation to some specific skill domains. These domains may differ importantly from the domain of philosophy, art, and other creative endeavours (and there are questions about the transfer of capacities from one skill domain to another), but I will leave this an open question, other than to say that I think there is a continuum here, a difference of degree rather than a difference in kind.

Of course, one needs to be able to adjust when the cricket bat is unready-to-hand, as Heidegger might say. When the Australian batsman Adam Gilchrist and Ricky Ponting were repeatedly dismissed by spin howlers on the subcontinent, for example, it is reasonable to think that for them to overcome their difficulties with these slow and turning pitches some kind of integration of reflection and practice is required, some kind of integration between acting and thinking, doing and knowing. Even if they would not want to be thinking too much during a test-match, their prior reflection and preparation will inform their procedural memory and help them to adjust. As such, we can and should complicate the Dreyfusian account a little. But the point is that phenomenology seems perfectly able to describe such experiences, as well as to explain the skill acquisition that is fundamental to such expertise. Phenomenological descriptions, for example, help us to see the manner in which our experience of time is aligned with the movements of a body-subject. They also help us to see the need for the adding of retentive and protentive elements to any idea of a now moment, rather than deploying a model of time that involves a series of instants. In addition, a phenomenology of bodily intentionality and anticipation helps to render explicable the ability of batsmen to respond in a timely fashion, since bodily know-how functions at a far quicker and more immediate way than would be suggested by the old representationalist model in which one passively perceives the sense data, then makes an active judgment regarding what to do, and then reacts, all while still attending to the trajectory of the ball.[17]

Of course, this is not all, or even a large part, of what phenomenological philosophers say about time, not considering the detailed descriptions of Heidegger on boredom and care, or Levinas' work in *Time and the Other*, to give two key examples. Moreover, I have not said much about various other phenomenological experiences of time, including reminiscence, nostalgia, nor the temporal differences between guilt and remorse. Instead, my focus has been on showing that the various temporal experiences we may have, which are associated with the past, present, and future, depend upon something like the structure of retention, primal impression and protention in Husserl, or what Heidegger called pragmatic temporality. As David Hoy puts it, "we experience ourselves as in time and as having a past, present, and future because our temporality involves the structure of protention, retention and primal impression."[18] Now, it is perhaps fair to say that the synthesis of time involved in what phenomenologists called the living-present involves emphasizing this integrative aspect, this gathering together, and in the work of Husserl and Merleau-Ponty it is also privileged. In the latter's work, our bodily intentionality aims to secure an equilibrium with the world and this provides a normalizing trajectory that allows us to succeed in various areas of expertise. The question is, however, whether this is but a phenomenological and/or psychological illusion as John Searle or Daniel Dennett might maintain, or a transcendental illusion as Deleuze might suggest, or whether there is something problematic about both of these kinds of dismissals of the phenomenological rendering of the times of our lives. While analytic philosophers often look to physics and the neurosciences for an answer to this question (the empirical conditions of objective time, and of our experience of time), for Deleuze there is no need to make this objectivist move: transcendental philosophy can reveal the partiality and, ultimately, illusory nature of this experience of the "now" and the living-present from within, rather than presupposing a view from nowhere. More specifically, Deleuze looks to what we might call a transcendental psychoanalysis. Transcendental philosophy and Freud are thought to get us beyond the time of consciousness and the time of embodied subjectivity, such that "I" become, as Widder puts it, a "multiplicity of subjects living different temporalities within the same not so unified being."[19] In a sense I am happy to accept this claim, and will indeed advocate that there are competing lived times of ethics and politics, as well as a normative aspect to the integrative experience of the living-present and that which breaks this living-present open and exposes it to the new and different. My own account of lived-time is indeed a pluralist one, but my worry with what I characterize as the poststructuralist paradigm is that the integrative time of the living-present is both treated as secondary (in a transcendental sense: we must look beyond the experience of the living-present to its conditions) and a strong ethico-political impetus is (one-sidedly) associated with that which disrupts.

Now it is true that much of the above account of the lived-time of the cricket player is indirect. It is also not entirely unfair to associate this with phenomenology more generally. But the question is what kind of direct image of time might be proffered instead, noting the long acknowledged aporias and difficulties with directly philosophizing about time, illustrated by Aristotle, Heidegger, Derrida, to mention a few, and famously lamented by Augustine. Assuming that one is not content to simply trace time from the empirical (that is, from post-Einsteinian physics and four-dimensionalism, where time's difference from space is ultimately effaced), is the solution to radically distinguish time from movement, bodies, etc. (as with Bergson) and to insist on time as a formal transcendental condition in a quasi-Kantian manner? (as with Deleuze). Such answers will certainly differ from the phenomenological accounts of time of (the later) Heidegger and Merleau-Ponty, where time and space (including movement) are intimately connected. Merleau-Ponty insists that one need not take the Bergsonian pill of radically separating time and space and privileging the former, and post *Being and Time* Heidegger also emphasizes time-space, and on certain interpretations, place.[20] For both, the metaphysical question of which came first, time or the subject/*Dasein*, is misplaced. As Merleau-Ponty says in *Phenomenology of Perception*, "We are not saying that time is for someone . . . we are saying that time is someone . . . We must understand time as the subject and the subject as time" (PP 422.) Given that we know that the subject for Merleau-Ponty is a body-subject, then motility and time are clearly bound up with one another on his view. Bodily motility and bodily intentionality seem from the beginning of life to be temporally tensed: proprioception is evident at the earliest stages of in utero life, and proprioception requires an experience of time and passage. Moreover, we have agreed with Shaun Gallagher that this results in a sort of prototype of self-consciousness, a condition for it, and this prototype of self-consciousness is indistinguishable from a primitive time-consciousness. In neonatal life, the perception of an object gives us hitherto undisclosed sides, sides that our attention might be directed toward and which we necessarily anticipate. If these kind of conditions of bodily subjectivity are also conditions of a "world" in Heidegger's sense, does this mean idealism? It depends on whether Merleau-Ponty means "time is the subject" metaphysically. If he did, this would seem to entail that so-called objective time is derivative of the time of our lives, and it might hence be protested what while our experience of objective time (objective time "for us" as Meillassoux might say) may be derivative of this lived time, that does not necessarily mean there is any metaphysical relationship of derivation. On this latter view, we might more plausibly interpret Merleau-Ponty's comment as simply referring to the manner in which our learning about the world is through and through temporally tensed, as well as the manner in which our situated experience of time is also what enables self-reflection and the constitution of subjectivity.

But for Deleuze time still needs to be unhinged from this too subjective a perspective. The key problem with this account of time is that it does not seem to offer an account of the advent of the new, and why it is that time (including our experience of it), is always cut, nicked, or broken up, and exposed to an unknown future. Deleuze's various books hence argue that genuine creativity requires a rather different experience of time, a form of time that the apparent excesses of sadism and masochism are more open to, and which he and Derrida both borrow from Hamlet, the Northern Prince, and call "time out of joint" (while some Deleuzians maintain that this transcendental condition cannot be experienced, there is much in *Difference and Repetition* and other texts that implies otherwise). On the other hand, it seems from my account of the cricket player that it is precisely the integrative aspects of temporal experience that open up a horizon in all of its difference and variability. This, of course, is one of Husserl's reasons for privileging the living-present: memorial time and narrative time in we which project particular futures all depend on this primary temporal immersion that is the living-present and are inconceivable without it. While Deleuze, Derrida, and others have given us sufficient reason to worry about this (idealistic) trajectory of grounding all aspects of temporality in the "living-present," any stronger claim than that, however, such as that the former are an illusion and tacitly a debased conservatism seem to me to be rather more tenuous. Moreover, whether their own transcendental philosophy of time is any better placed than Husserl's is not so clear, being generally illuminative rather than strictly necessitarian, as I will suggest in the remainder of this book. In addition, while a normative emphasis on the nick of time—the *contretemps*—that sunders identity has value, we have also seen that there are normative virtues associated with the temporal amnesia that I have described. As Nietzsche says, "without forgetfulness, there can be no happiness, no hope, no present."[21] Let us briefly consider Deleuze and Derrida on "time out of joint" before returning to such issues.

Deleuze and Derrida: Time Out of Joint

While Deleuze's account of time out of joint is argued to be a formal condition for experience to have the structure that it does, it is important to note that it is not merely a neutral transcendental claim, also being explicitly associated with actual traumatic experiences like those of Hamlet (or learning to swim, etc.) and having a clear normative register. When Shakespeare's Hamlet declares that time is out of joint, or unhinged,[22] this is predominantly because various actual events have monstrously violated Hamlet's sense of his world, including most notably the murder of his father, and his mother's remarriage to his uncle. Such events, of course, famously paralyse him for some time before he ends his prevarication and becomes equal to the act. And this kind of aporetic impasse, this undecidability, is essential to the notion of "time out of joint" for Derrida in *Spectres of Marx*. Indeed, Derrida's point—which hovers between being a form of conceptual analysis and being more metaphysically committed—is that the only time in which something can ever happen, is when time is out of joint, when there is a constitutive not knowing what to do and how to be. "Time out of joint" hence has something to do with what Derrida calls *contretemps*, which indicates an unforeseen occurrence or event, but is more literally translated as the untimely, or counter-time.[23] Derrida is clear that the untimely is not, however, atemporal, but is rather counter to linear time, and teleological history, with its seasons, regularity and order. *Contretemps* is the condition for vulgar time, the time of the present, but it is also that which breaks that living-present apart. It is the time of the event.

For both Derrida and Deleuze, there is a sense in which this disjointed experience of time is more pervasive than that being a response to actual worldly trauma, but is also a condition for our experience of the living-present, albeit one that is covered over and concealed by experience itself. In this respect, time out of joint might refer to the manner in which waiting is essential to all experience, as well as the manner in which every experience contains an aspect of lateness.[24] In *Spectres of Marx*, Derrida calls this kind of relation to time "anachronism." Anachronism is an error of sorts, a relating to an event or custom or ritual as if from the wrong time. It suggests someone or something is out of harmony with time, the living-present. While there are clearly forms of anachronism that may be problematic—interpreting the past from the perspective of our own current predilections and interests—there is also something positive to anachronism for Derrida. In times of crisis when the new (and potentially violent) threatens to erupt in revolutionary crisis, Derrida suggests the more that one needs to borrow from past, and to attend to spectres and hauntings.[25] This might not be in the form of nostalgia for the past, but some kind of spectral or uncanny visitation is required. Of course, precisely what we saw the sportsperson does not seem to want, or need, is such visitations.

In arguably the central part of *Difference and Repetition*, the account of the eternal return of difference and the disjunctive synthesis of time, Deleuze also invokes the Northern Prince. For Deleuze, prior to his father's murder, Hamlet's experience of time was oriented around "those properly cardinal points through which pass the periodic movements which it measures"—time was measured in relation to orderly movements of the world, sun and moon, dinner, duties, etc. In contrast to this, Deleuze says, "a time out of joint means demented time or time outside the curve . . . freed from the event which made up its content, its relation to movement overturned" (DR 88). The movements by which time had been measured are disrupted, leaving only an empty form of time that eschews the unity of the subject. Widder pays a lot of attention to this fractured self in his *Reflections on Time and Politics*, explaining its psychoanalytic provenance and seeking to problematize the normalizing trajectory of bodies seeking an equilibrium (the return of the same rather than difference, we must assume). While neither he nor Deleuze want to dispute that this happens (at a superficial level), they want to revalue another kind of temporal condition for the living-present, which is also both an experience of sorts, hence the analogy with Hamlet, as well as a kind of regulative ideal for how to live. This unhinging, to return to Shakespeare, fractures the self and opens it to a becoming-other of some kind. On Widder's view, time is a structure or disjunctive synthesis that ungrounds movement, including the idea of the flow, or passage, of time. As he puts it: "'time' names the structure, not the measure, of change. It is a kind of being out of synch with oneself that is the condition of anything to change or move."[26]

Not only is time out of joint a transcendental condition for all experience, but Deleuze also indicates that we can better affirm and embrace this time, if only we could become good throwers of the dice, embracing both chance and necessity. Previous chapters have explored the manner in which this is so in *Difference and Repetition* and *The Logic of Sense* respectively. In both texts, Deleuze valorizes those learning experiences that force us out of any such equilibrium with our environment; the kind of structural coupling between subject and world that is pivotal to the constitution of a living-present. Such analyses do have a certain phenomenological resonance, even if they are also argued to be more than that. We might, for example, invoke a related image of the sportsplayer who "counter-actualizes" situations, and for whom it is a matter of not simply responding to the actual stimuli (even in the attenuated sense of actuality with its projective and retentive aspects as described previously), but is fundamentally about negotiating the intensities provoked by past experiences and their hopes for an unknown future. The famous Deleuzian war-time example is Joë Bousquet in *The Logic of Sense*, and in *A Thousand Plateaus* Deleuze and Guattari claim that nomads change their habits so as not to change their habitat (ice and desert), but migrants move and change their habitat so as not to change their habits.[27] In relation to this opposition, Deleuze and Guattari side with the nomads as a regulative ideal at least—they call them the "noumena of history"—but perhaps a philosopher like Merleau-Ponty and cognitive scientists indebted to him, are tacitly on the side of the migrants. Although circumstance changes, and we must consistently adjust, there is a normative impetus to attaining maximum grip on an environment (this is an evolutionary pressure too), to what cognitive scientists call structural coupling between organism and environment. Deleuze and Guattari argue that any genuine creativity or learning must be provoked by something traumatic, or at least the possibility of trauma must be omnipresent to sustain creative performance in any domain, but perhaps particularly philosophy and art. It is easy to get an intuitive grasp of what they are on about in this regard. We have all seen great performers in the early stages of their careers, who, some short time later, flush with success, are totally confident in coping with the pressures of live performance, but have lost something vital from their performance. And Deleuze, Widder, and others, are clearly right to suggest that life is not exhausted by bodily coping (and the time of the living-present, *l'habitude*), that even the activity of the cricket player is not done justice to without some reference to what we might summarize as time out of joint, in both its formal and also more experiential guises. Perhaps there are also some strategic reasons for privileging "lost time" in modernity, given the sense in which the pragmatics of worldly time, to return to Heidegger, are increasingly dominated by ordinary time, clock-time. But worlds and lives change, whether the self is fractured, whether Joe is thrown into the volcano or not. If there is a law, it is that of change, but it is never clear to me that the transcendental arguments about time out of joint (or structural equivalents) that are bound up with a recognition of this fact are compelling. They typically depend upon an opposition between the event and inexorable sameness, predictable predicates, and so on. But is every philosophy of mediation, of continuums, necessarily condemned to be unable to explain the event/change? It is not clear that this is so. Transcendental reasoning of this sort depends upon a contrast that excludes other possibilities and cannot establish that its alleged conditions are the uniquely valid ones. As such, it seems to me that what we are witness to in these temporal disputes between phenomenologists and poststructuralists is an account of the time(s) of our lives that is irremediably split in both of these directions, and which problematizes any attempt to adequately ground the one in the other. In regard to the association between time and politics, this quasi-transcendental necessity also precludes any too easy decision in the realm of the ethico-political (e.g., change and rupture vs. sameness and coping). The remaining chapters will attempt to further justify this claim, and exhibit the manner in which Deleuze and Derrida (to a lesser extent) betray this quasi-transcendental necessity despite themselves often glimpsing it in their own philosophies.

10

Dreyfus, Merleau-Ponty, and Deleuze on *l'Habitude*, Coping, and Trauma in Skill Acquisition[1]

> The state of a moral man is one of peace and tranquillity, while the state of immorality is one of perpetual unrest.
> (Marquis de Sade)

One of the more important and under-thematized philosophical disputes in contemporary European philosophy today pertains to the significance that is given to the interrelated phenomena of habituality, skillful coping, and learning. This chapter attempts to make this largely unacknowledged dispute explicit, by focusing on the work of the Merleau-Ponty and Heidegger-inspired phenomenologist, Dreyfus, and contrasting his analyses with the poststructuralist position on these issues, as it is exemplified in the work of Deleuze, particularly in his key text, *Difference and Repetition*. Both Deleuze and Dreyfus pay a lot of attention to learning and coping, while arriving at distinct conclusions about these phenomena with a quite different ethico-political force. While Dreyfus' phenomenology prioritizes the various coping techniques that lead to relative equanimity and equilibrium within a given environment (and this carries implied consequences for morality along the lines of the above epigram that Dreyfus makes explicit), there is a very different move in Deleuze's understanding of coping and learning. Deleuze valorizes those learning experiences that force us out of any such equilibrium with our environment, and he implicitly denigrates the worth of the kind of embodied and habitual coping techniques that Dreyfus thinks are so important. Within certain bounds (the risk of what Deleuze calls chaos in his collaborative writings with Guattari), we must aspire to be the perpetual apprentice, to encounter new situations that are inassimilable to our average coping techniques, and to become a nomad who is never at home.[2] These are very different ethico-political injunctions and the question of the role of habit and skilful coping in learning is vital to both of them. By getting to the bottom of the latter, my hope is to problematize certain aspects of the former in the work of both philosophers. In Deleuze's case, it will be argued that despite the acuity of his philosophy of difference he nevertheless adopts a problematic position on learning that will be termed "empirico-romanticism." While I will agree with the general thrust of Dreyfus' foregrounding of habit and skilfull coping, even in the political realm, there are also some risks associated with his view (notably of it devolving into a conservative communitarianism) that this juxtaposition with Deleuze helps to illuminate.

Dreyfus and Skilful Coping

One of the foremost interpreters of the work of Heidegger and Merleau-Ponty in the American continental tradition, Dreyfus also uses their respective philosophical resources for his own purposes—hence the nomenclature "Dreydegger," bestowed upon him by John Searle. In particular, he draws on Heidegger's insistence in *Being and Time* upon the priority of coping and the practical, ready-to-hand way in which we most primordially engage with the world, as well as on Merleau-Ponty's intricate discussions of embodied habit in *Phenomenology of Perception*.[3] While perhaps most famous for the criticisms that he has launched against cognitive science and the possibility of artificial intelligence, ever since these two disciplines emerged, throughout his work has been concerned with learning, habituality (or what he prefers to call *l'habitude* because the French terms resists some of the connotations of mere habit that have accrued in English), coping, and skill development. In various different essays and books, and often with his brother, Stuart, he traces the different modalities of skill development from the beginner through to the expert and argues that we need to take a fresh look at what expertise consists in. His studies convincingly show that cognitive and reflective calculations play less of a role in the activities of a master chess player, expert car driver, or basketballer than we usually assume. While deliberate and calculative rationality are more central to the activities of the beginner and the competent as they try and decide on their next course of action, this is not the case for the expert who can generally perform exceedingly well without having the time to think, plan, or even deliberately intend a certain outcome, precisely because their coping techniques have been honed to such a degree that they are solicited by the situation to respond in ways appropriate to the particular circumstance (this is not just a matter of internalizing and following rules, as we will see). Through habit and skill, they have acclimatized to the situation and all of the nuanced possibilities opened up by it, such that they can just go with the flow—in Dreyfus' own favored example, they are in what Larry Bird and numerous other sportsmen and women have called "the zone," where things just happen and can only retrospectively be understood as deliberative. Of course, sometimes things will go wrong, and calculative reflection will be necessary, but this is rarer than we have traditionally thought.

If correct, this understanding of skill development constitutes a refutation of much of the view of learning that runs from Plato all the way through to Piaget and Chomsky, which argues that a beginner starts from specific singular cases about which they know very little, and, as they become more and more proficient in a particular activity (through practice and reflection), then abstracts and interiorizes more and more sophisticated rules. For Dreyfus, it is true that as we gain in experience in a particular activity we do encounter more and more solicitations to respond and to act, but these are not general rules. Rather, they are more and more differentiated and refined (i.e., singular) solicitations to act. Skill acquisition moves in the opposite direction to what is usually assumed: from the representation of abstract rules as a beginner and competent practitioner (and calculative thinking on behalf of an agent as to the best employment of particular rules), to a kind of immediate and habitual attentiveness to particular cases and circumstances as an expert. This is why a master chess player can continue to beat the merely good player, even if they have to respond immediately and while counting or doing something else that preoccupies their cognitive capacities, as one of Dreyfus' studies shows.[4] It is this kind of embodied engagement with the world that is fundamental to learning and which Dreyfus insists that computers (which are calculative and representational) will not be able to emulate adequately.[5]

It should also be noted, though, that such spontaneous receptivity to an environment is not solely the province of experts, and nor does it apply only to the sporting realm. Rather, it is something that all of us have, in all of our endeavors, to some extent. Indeed, Dreyfus insists upon the way in which our daily activities and choices are facilitated by this embodied coping that is prior to reflection and that does not require any cognitive representation of its goal. In other words, this primordial skilfull coping both makes possible the acquisition of rules and the cognitive itself, and also contains within it the end or telos of embodied life, which is, to put it bluntly, minimizing tension. For Dreyfus, it is important not only to distinguish these kind of practical, habitual, and yet flexible responses to a situation from more reflective, deliberate and calculative behavior, but also from what might be denigrated as the instinctual and deterministic repetition of past behaviors. His argument that coping is not reducible to instinct seems to be supported by a recent study by animal behaviorists at Oxford University that attempted to determine precisely how it is that homing pigeons manage to find their way home. While they have an instinct to find their way home devoid of geographical measures, and this is how they navigate over the sea, in practice, where possible, tracking devices have shown that the pigeons actually follow roads or other key landmarks, even if it sometimes involves flying considerably longer distances—the phrase "as the crow flies" is hence misleading, at least if we can assume that crows have a similar behavioral disposition to pigeons. Just as when we walk home we usually do so habitually rather than ponder and reflect upon the best route to take, pigeons also "home in" on a habitual journey. It looks like it is somehow less stressful for this bird to fly down a road and follow it right, than to use their in-built navigational instincts.[6]

In explaining this kind of behavior, Dreyfus draws on several key ideas from Merleau-Ponty's *Phenomenology of Perception*, including the notion of maximum grip, which names the body's tendency to respond "to solicitations to act in such a way as to bring the current situation closer to the ... sense of an optimal gestalt"[7]—that is, to negotiate an equilibrium with one's environmental and material situation. While this kind of ongoing effort at minimizing stress might not be said to be either a case of learning or a skill, for Dreyfus it is the basis upon which learning and skill development can occur. He employs Merleau-Ponty's concept of an "intentional arc" to help explicate skill development, suggesting that it "names the tight connection between the agent and the world, viz. that as the agent acquires skills, those skills are 'stored', not as representations in the mind, but as dispositions to respond to the solicitations of situations in the world." It is the link that the body establishes between action and perception, and Merleau-Ponty suggests that it subtends the life of an individual consciousness and "projects round about us our past, our future, our human setting, our physical, ideological and moral situation" (PP 136). Dreyfus also wants to insist, again inspired by Merleau-Ponty and Heidegger, that this mode of behavior points to a middle-way between the emphasis on reflection and cognition that it is at work in intellectualist accounts of behavior (the camp that the modern cognitive scientist usually fits into), as well as in empiricist accounts of learning and behavior (in this respect, one important question for us to address will be whether or not Deleuze's reconfigured empiricism problematizes or falls outside of this understanding of empiricism). Dreyfus' deeper philosophical claim is that rules and cognitive thinking per se are parasitic upon this more practical and primordial engagement with the world evinced by terms like maximum grip, *l'habitude*, coping, and the maintenance of intentional arcs.

Merleau-Ponty on Schneider and the Pointing/Grasping Distinction

In order to understand this claim that cognition is dependent upon a more primary mode of engagement with the world that is noncognitive and best evinced by an attitude of "I can" rather than "I think," it is helpful to consider

Merleau-Ponty's attempts to distinguish the unreflective grasping body from the reflective pointing one. At various different points in *Phenomenology of Perception*, Merleau-Ponty discusses a patient by the name of Schneider who on account of brain damage sustained during a war is unable to perform bodily movements in the normal flowing manner (PP 103-110). For example, while he can blow his nose if he feels that need, or *grasp his nose* if it has been bitten by a mosquitoe, he cannot perform equivalent actions if his eyes are shut and he also cannot *point to his nose* if asked to. If he watches his hand, eventually he can guide it to his nose, but it is a laborious and awkward process. These kind of abstract and reflective activities go through the inter-mediary of the objective world and Schneider cannot perform movements that are not a response to an actual, present situation. When attempting to point, Schneider is like a distanced observer to himself. In fact, he appears as a paradigm case of what it would be like to have a mind that ordered the body to do things and we can hence practically see just how problematic this model is. Schneider does treat his body as such an object and it makes actions like pointing to his nose almost impossible. However, when confronted with a practical task such as cutting paper, Schneider does not have to locate his hands before moving them; the scissors on the table and the task of cutting paper immediately and unreflectively mobilize potential actions and solicit him to react to them in certain ways. Concrete movements and acts of grasp-ing, such as the blowing of his nose, enjoy a privileged position for him and, Merleau-Ponty argues, for all of us; it is the main example of what he calls motor intentionality. The phenomenological distinction being drawn here is that a grasping movement is from the outset at its end, in that it has an inten-tionality that means that success depends upon achieving a certain end (the act must be successfully completed in the real world, i.e., actually grasping the coffee mug on the table), whereas this is not necessarily the case for the more abstract activity of pointing. Grasping also relies more heavily upon anticipation, and the understanding of space involved in it largely resists our explicit thematization and understanding. Whereas demonstrative pointing depends upon knowledge of the object that differentiates it from all of other objects, or in Gareth Evans' terms involves "knowledge of the object's objec-tive location, not just knowledge of its egocentric location,"[8] the intentional content of "grasping" perceptions and actions is "prepredicative," not ca-pable of being understood in abstraction or divorced from the activity itself. Rather, it is a practical, embodied knowledge that discloses the world to us but cannot itself be captured in the process of doing so.

    In response to this case, we might follow Merleau-Ponty and ask the obvious question: "If I know where my nose is when it is a matter of hold-ing it, how can I not know where it is when it is a matter of pointing to it?" He suggests that the problem is that Schneider either has an ideal formula for a particular movement that he works out in his head before acting, or he launches blindly into movement. There is no feedback between these two very different attitudes, whereas for most of us active movement is indissolubly both movement and consciousness of movement and we presuppose a mutual presence of body and object in our pointing. Merleau-Ponty is also keen to point out that Schneider's difficulties cannot be accounted for on the basis of either empiricism or intellectualism and Martin Dillon summarizes his point as follows: "The patient cannot be understood as suffering from a purely physical disability (as empiricism would have it) because he can perform the physical movements; but neither can he be incapacitated in a purely psychical way (as intellectualism claims) because he can understand the goals to be achieved."[9] Dreyfus and Sean Kelly use related arguments to trouble contemporary cognitivist and empiricist accounts of behavior, suggesting that skilfull motor behavior, such as the grasping of a mug, cannot be accounted for by the non-intentional purely reflex act (it is different from an instinct), nor in terms of any kind of rational cognitive processes (there are not necessarily any going on). As Kelly puts it, "the empiricist account fails because its purely muscular vocabulary doesn't allow it to distinguish between mere reflex movements and directed, skilful motor actions. On the other hand, the cognitivist account fails because its purely cognitive vocabulary doesn't allow it to distinguish between unpremeditated motor actions like grasping an object, and premeditated, cognitive actions like pointing at one."[10] For both Kelly and Merleau-Ponty then, pointing and grasping are based on two different phenomenological understandings of space and place that cannot be ignored in any coherent account of embodied learning and coping. The former is calculative and rational, the latter is precalculative and prepredicative. However, the further claim made by Merleau-Ponty, Heidegger, and Dreyfus alike (although Kelly is not so sure about this[11]) is that grasping has an ontological priority, although that is not to rule out or denigrate pointing and calculating. One way of justifying this claim is to note that the understanding evinced in grasping can be independent of the understanding involved in pointing, but the reverse does not seem to apply. Schneider can grasp, but he cannot point, for example, but there seem to be no recognized occasions where the reverse obtains and someone can point but cannot grasp. On this view, grasping is then the condition of possibility of pointing, and this kind of practical noncognitive grasping is, for Dreyfus, Kelly, and Merleau-Ponty, largely what takes place in learning. For Deleuze though, this priority given to grasping and the "I can" is called into question, replaced by a transcendental priority accorded to the different, the future, and the disruptive. In Derrida's work there are also consistent refrains against this "I can" and practical coping with our environment. In an interesting passage in *Aporias*, he comments on the priority that Heidegger gives to coping and suggests that it emphasizes possibility in two senses, the second of which is described as, "the sense of ability, of the possible as that of which I am capable, that for which I have the power, the ability or the potentiality" (A 62). But to return to Deleuze, his argument, will be that one can only repeat—i.e., grasp, behave habitually—because of difference. While such an analysis may be cogent on the transcendental level, albeit somewhat one-sided as we will see, I will level at least two further objections to it: first, it is not clear that it is necessarily in opposition to Dreyfus and Merleau-Ponty's position (and Deleuze's own position implies that it is thus opposed, in the distance that his work continually insinuates between coping and learning); and second, nor does it entail the overstated normative priority that Deleuze accords to the empirical experiences of apprenticeship and trauma in learning.

Morality as a Skill? Phenomenology as Communitarian?

Before getting to the work of Deleuze, however, we need to return to the epigram from the Marquis de Sade with which we began this chapter: "the state of a moral man is one of peace and tranquillity, while the state of immorality is one of perpetual unrest." It might still be unclear exactly what relation Dreyfus' analyses of embodied equilibrium have for morality and hence the quotation in question. But not only does Dreyfus' analysis of skill acquisition have clear normative implications, but he explicitly argues for an isomorphic position regarding moral maturity, suggesting that it is primarily about an ethical comportment to situations in the world rather that about coming to have more sophisticated cognitions and judgments about principles and rule-following.[12] For Dreyfus, in the stage of moral competence we may try and figure out if our behavior is suitably reflective of, say the golden rule that one should "do unto others as they have done unto you," or Kant's categorical imperative, which asks us to imagine the consequences of the universalization of our actions. On his view, however, if we are lucky enough to attain moral maturity, we generally will not reason about rules and principles, or even act deliberately, but will simply spontaneously do what has normally worked, just as we saw was the case with expert chess playing.

There are, of course, a few factors that trouble any extension of Dreyfus' analysis of skill acquisition to the moral domain, as he admits. Most notably, with chess-playing and driving, for example, we get immediate feedback as to the adequacy or inadequacy of one's actions, yet with ethical behavior we don't necessarily get this kind of practical reinforcement. Our actions in the world may induce smiles from our colleagues at work, the odd "thank you," and sometimes even genuine engagement that transforms the self-understanding of the people involved, but the feedback loop, and therefore our ability to form intentional arcs with the environment that we are acting in, is not as immediate and clear as it is in the kind of examples that Dreyfus generally focuses on. Similarly, in relation to the grasping and pointing distinction, we can see that were we to attribute something akin to a kind of foundational "grasping" to expert moral activity (Dreyfus does and calls it intuition), it is again the case that moral decisions and actions are not immediately at their end in the way that we have seen that the activity of grasping a mug is inextricable from the projected successful grasping of it.

That said, there is a type of feedback that might be enough to provide the foundation for an immanent account of the genesis of moral maturity. Rather than rationality being privileged as the key to establishing an objective moral code, we might instead see that moral maturity is acquired over a period of time through the way in which bodies increase in capacity and we might even say power, in the Spinozian sense, to the extent that their capacities to affect and be affected become more developed and differentiated. The key question, of course, is whether there is a dulling of such capacities in certain forms of repetition (such as coping), or whether such repetition is a condition for the enhancement of these capacities. For now, we might simply note that all actions have consequences in the world at large, and the affective states that accompany these actions will depend at least partly upon how they are received by other people, which will in turn react back upon the body's capacity to act. As Paul Patton observes, interestingly enough in his book on Deleuze, "to the extent that these actions are successful, the feeling of power will be enhanced: to the extent that they fail, the feeling of power will be depleted. In turn, these affective states which accompany actions will react upon the agent's capacity to act. In other words, there is a feedback loop between the success or otherwise of attempts to act and the agent's capacity to act."[13] It seems then, that Dreyfus' immanent and nonreflective account of the genesis of moral maturity bears at least some relationship to the Nietzschean and Spinozian account of morality that Deleuze explicitly endorses in texts like *Nietzsche and Philosophy* and that remains a major factor in *Difference and Repetition*.

If Dreyfus' analysis of "moral skills" is correct it constitutes an ambitious challenge to Kantianism in most of its contemporary guises and disguises, including the contemporary positions of Rawls and Habermas. In fact, Dreyfus' phenomenology of coping and skill acquisition lends itself to a kind of communitarianism (and here it parts company with Deleuze),[14] and, in places, a kind of virtue ethics. Dreyfus is not the only phenomenologist whose work alludes to this connection. Consider the work of neophenomenologists like Charles Taylor, Paul Ricoeur, and others, all of whom have been influenced by Heidegger. Despite their apparent left-wing radicality and

Marxist leanings, it is also worth recalling the early explicit critiques of liberalism advanced by existential phenomenologists like Merleau-Ponty and de Beauvoir.[15] They are essentially communitarian in their focus upon the difficulties inherent in liberalism's atomistic conception of the subject; they will not accept the supposition of a rational disengaged agent, even as a regulative ideal. This link between phenomenology and forms of communitarianism is perhaps not as surprising as it initially appears.[16] After all, the question of whether or not there is an ontological and normative priority to be accorded to a nonreflective coping with a situation (with which most post-Husserlian phenomenologists agree) links up with the debate between liberals and communitarians on the question of the preconditions of agency and individuality. It is even arguable, although I will not pursue this here, that large parts of the whole liberal/communitarian debate can be reconceived as an isomorphic variation on the pointing/grasping debate, with its key distinction between an abstract mode of rationality and a corresponding form of agency, and a more practical immersion in a situation that precedes agency.

But the key point is that, for Dreyfus (as for virtue ethicists), there is not likely to be any overall theory or principle that unifies the behavior of someone whom we consider to be morally mature. Instead, such a person attends to the particularity of each situation, but can only do so because of the maintenance of *l'habitude* and intentional arcs that allow us to establish links with our environment (and hence to create a world in Heidegger's sense), such that we are solicited to respond to it in more and more refined ways. Dreyfus' further claim is that the search for rational principles that undergird and unify any such activity in fact blocks further development beyond the stage of moral competence. While he admits that deliberation is sometimes necessary and helpful (and can even be a type of expertise), to read it as endemic to morality, or the telos of morality, in the manner of the early Rawls and Habermas, is ultimately unjustifiable. Rather than see critical rationality as overriding our intuitive ethical comportment, Dreyfus advocates a reversal of this priority and sides with caring over justice in the age-old debate. On his view, considerations of justice tend to be part of the competent stage of skill acquisition, but something like practical wisdom is more typical of the phenomenology of expertise. He suggests that while one aspect of moral life, and most of moral philosophy, has been concerned with choice, responsibility, and the justification of validity claims, we should instead take more seriously the idea that moral maturity is better understood in terms of a skilled ethical comportment, "which consists in unreflective, egoless responses to the current interpersonal situation."[17] Any investigation of ethical experience should begin on the level of spontaneous coping and he suggests three general methodological precautions for all moral philosophy:

1. We should begin by describing our everyday, ongoing ethical coping.
2. We should then seek to determine under which conditions deliberation and choice occur.
3. We should beware of making the typical philosophical mistake of reading the structure of deliberation and choice back into our account of everyday coping.

It seems that on all three counts, most versions of liberalism, and certainly the Rawlsian liberalism of *A Theory of Justice*, fails.[18] In fact, arguably all theories of rational choice, including game theory and decision theory, make this mistake of privileging the structure of deliberation and choice, both in regard to issues to do with distributive justice and individual morality.

Moreover, it is interesting to note that the poststructuralist reservations about the political philosophy of Rawls, Habermas, and others are *prima facie* rather similar to those expressed here by Dreyfus. The poststructuralists are also highly concerned with the privilege accorded to a universalizing rationality and the lack of attention paid to the aporias and limitations incumbent upon any and all such perspectives. They pay more attention to singularity, difference, that which resists being absorbed by a programmatic set of rules, or is between a dominant sets of rules, as in Lyotard's discussions of the *differend* and Derrida's preoccupation with undecidability. At the same time, we must also note an important difference between them at the outset. Dreyfus' position intimates that the experience of trauma and discontinuity (what Derrida might call the aporetic) are confined, more often than not, to beginners and competent practitioners, and even then, are rare, because of our omnipresent embodied intentionality that is always-already adjusting us to the peculiarities of our situation. For Dreyfus, the sort of problems that require reflection, decision, and, perhaps more pointedly, that involve radically disruptive affects, normally do not arise and are less likely to arise given greater familiarity and expertise within a particular milieu. It follows from this that on his view moral maturity results in having fewer problems and a more sturdy equilibrium.

While Deleuze would agree with the pragmatist and nonrepresentational intentions of Dreyfus' work,[19] he would be concerned with the tacit privileging of this embodied intentionality that seeks the diminishment of problems. After all, as Deleuze frequently suggests, "something provokes us to think": that something being affective problems and paradoxes, the latter of which is important because it displays a difference that cannot be absorbed or totalized within a common element. For Deleuze, problems are precisely what gives life value by impelling and motivating us. They are "inseparable from a power of decision, a *fiat* which, when we are infused by it, makes us semi-divine beings," (DR 197) and the solution, such as it is, is about mobilizing new concepts, not dissolving them in a certain Wittgensteinian fashion. In this respect at least, there is an important difference between Deleuze and Dreyfus, and it seems fair to suggest, preliminarily, that although the everyday pragmatics of coping are important, and politically important, if it entails or presupposes that life is easy and comfortable, then there is reason for reservation. Were these coping techniques that habituate us to the world so refined that we were to almost absorb the world within a radically enlarged body-schema, then the intentional arcs established through skills would clearly be drawn implausibly tight. It would be akin to saying that we and the world coincide, thereby ignoring and downplaying all that is surprising and traumatic. While neither Dreyfus nor Merleau-Ponty are necessarily committed to this, there does seem to be an element of what Deleuze would denigrate as the ur-doxic desire for a harmony between self and world in their work.[20] At the same time, it is also based in some rather convincing analyses that highlight the inevitability of, at the very least, a strong coimplication of self and world. We can begin to get a better grip on these difficult questions by considering Deleuze's own analysis of habit and learning, paying particular attention to the distinction that he wishes to draw between them.

Deleuze and Habit

Habit is very important to Deleuze's philosophical position in *Difference and Repetition*, being the synthesis of time that makes possible the living-present and subjectivity. However, as we saw in chapter 5 it is clear from his analysis that something like the time of the future (the time of the eternal return of *difference*, where subjectivity and habit expressly do not return) is regarded as evincing a more fundamental philosophical difference and as a morally preferable exemplar for how to live. This is because, for Deleuze, habit is the foundation of what he calls and denigrates as "good sense," which refers to the dominant supposition that things move from the unpredictable to the predictable. There is a temporality betrothed to this good sense, in that it understands the relationship of the past to the future as paralleling, and perhaps even equivalent to, the movement from the particular to the general. It is habit that makes possible the tacit assumption of good sense, which, along with what he calls common sense, is the key to explaining all kinds of social and philosophical dogmatisms. For Deleuze, the problem with habit is that it generalizes and abstracts from the singular points of experience, because, as he says, in it:

> we find two major orders: that of *resemblance*, in the variable conformity of the elements of action with a given model in so far as the habit has not been acquired; and that of *equivalence*, with the equality of the elements of

action in different situations once the habit has been acquired. As a result habit never gives rise to true repetition: sometimes the action changes and is perfected while the intention remains constant; sometimes the action remains the same in different contexts and with different intentions (DR 4-5).

Deleuze goes on to say that *if* repetition is possible (meaning the repetition of genuine difference and it would be hard to think that this question is really bracketed away), it is between or beneath the two generalities of perfection and integration and the platitudes of good sense that habit grounds. It is immediately apparent that both perfection and integration *do* feature quite heavily in Dreyfus' analysis, the former in the normative impetus given to mastery and expertise, and the latter in the idea of establishing an equilibrium and minimizing any kind of embodied tension. That said, many questions still remain unanswered, notably whether Deleuze's analyses of coping and learning are phenomenologically accurate, because we are still going to have to grapple seriously with Dreyfus' claim that these two things, expertise and equilibrium, rather than giving us the order of generality, are actually what makes possible true attentiveness to the particularities of a situation. We at least seem to have seen that what goes in these more highly refined and attenuated coping techniques (i.e., skills) is not about mere generality. For Dreyfus, it gives us more specifically nuanced singularities and hence allows for particularity in a way that does not make the empiricist mistake of assuming the existence of some kind of blank slate, or, with a provocative view to some questions that we will pose to Deleuze, nostalgically desiring a return to such a tabula rasa (even if it is acknowledged that it could never have actually existed).

Deleuze and Learning

In his influential account of learning, Deleuze resists any suggestion that learning can be adequately comprehended within the realms of the phenomenological living-present that copes with and inhabits its surroundings. In fact, he tells us that there are two kinds of repetition at work in learning: first, the habitual one of everyday coping; and second, the more radical repetition of differences, pure contingency (pure time), which is in fact what makes possible the generality of habit. He goes on to say that these two repetitions are not independent, but suggests that the latter repetition is "the spirit of every repetition … it forms the essence of that in which every repetition consists: difference without a concept, non-mediated difference" (DR 25). Without yet considering the ethico-political impetus that his work draws from the experience of traumatic learning and these singular differences that are said to undergird habit, on the transcendental level there is much

to be said for this philosophy of difference. Nor is it necessarily antithetical to Dreyfus' own analysis, which acknowledges that habit on its own is
not sufficient to explain learning and skill development (the establishment of
something like an intentional arc is required for that). Indeed, on first glance
there are some important similarities between their respective analyses. For
Deleuze, as for Dreyfus, learning is nonrepresentational and he too stridently
distances himself from rule-following understandings of learning that depend
upon calculative reflection and knowledge. Instead, Deleuze contrasts "do
as I do" teaching with "do with me" teaching (much as the work of the later
Heidegger does), the latter of which he endorses because it offers signs to
be developed in heterogeneity rather than being simply gestures for us to
reproduce or imitate (DR 23). For him, "here as elsewhere, becoming conscious counts for little" (DR 19), and he suggests that learning expresses
"that extra-propositional or sub-representative problematic instance: the presentation of the unconscious, not the representation of consciousness" (DR
192). While there are important questions about the connections that obtain
between habit and the unconscious that cannot be addressed here, Deleuze
goes on to add that this kind of extra-propositional encounter that goes on in
learning, in which thought finds within itself something that it cannot think,
"is incomprehensible only from the point of view of a common sense, or that
of an exercise traced from the empirical" (DR 192).[21] What Deleuze calls
common sense, with its privileging of recognition, is closely aligned with
what Merleau-Ponty and Dreyfus criticize as intellectualism, which refers to
any of the various philosophical positions (including idealism, rationalism,
and contemporary cognitivism) that privilege the meaning-bestowing role of
the mind. Deleuze also rejects, if not empiricism *tout court* (as he claims to
find the hidden secret of this tradition in *Empiricism and Subjectivity*), then at
least what he would call crude forms of empiricism. In the work of Dreyfus
and Merleau-Ponty we have seen that a similar dualistic alternative is also
false: neither empiricism nor intellectualism can understand the indetermi-
nacies of perception and the ambiguities of embodied life, particularly the
phenomena of habit and grasping, both of which are prepredicative. Indeed,
in this context it is helpful to recall one of Merleau-Ponty's own comments
about skilful habit, which seems quite closely related to Deleuze's understanding of learning. He suggests, "if habit is neither a form of knowledge nor
an involuntary action, what is it then?" According to him, "it is knowledge in
the hands, which is forthcoming only when bodily effort is made, and cannot
be formulated in detachment from that effort" (PP 144). For both Deleuze and
Merleau-Ponty, there are important philosophical lessons to take from this
apparently basic phenomenon. Intermediate between knowledge and non-
knowledge, Deleuze suggests that it is "from learning, not from knowledge,
that the transcendental conditions of thought must be drawn," something that

Merleau-Ponty agrees with (DR 161; PP 61). Moreover, it almost recapitulates the basic methodological precautions that Dreyfus offered us in regard to moral philosophy—that is, to begin with the everyday coping and learning of skills, then draw the conditions of thought from that analysis rather than artificially importing considerations of rationality into this domain. Deleuze goes on to suggest that with this recognition of the transcendental import of learning, a more radical understanding of time is also introduced into thought, the implication being that mere considerations of knowledge are inevitably atemporal (in this respect, recall the preoccupation in contemporary political philosophy with issues to with distributive justice, which tend to be "time-slice" theories, tacitly predicated on the assumption that goods and resources descended ahistorically from heaven only to be synchronically examined and distributed). However, Dreyfus' embodied coping techniques are likewise prior to epistemological concerns, and while they may be said to legitimate the living-present by Deleuze, Derrida, and others, and hence not to truly introduce time into thought in the way that they promote, there is more going on in Dreyfus' analyses than such a formulation can recognize. Rather than being merely a reactive force, tantamount to slave morality, such coping techniques simultaneously carry the weight of past sedimentation and yet are also productive of a world of radically differentiated possibilities that may be anticipated but never epistemologically known, as becomes most apparent in Spinosa, Flores, and Dreyfus' *Disclosing New Worlds*,[22] and as we saw in the previous chapter's extended discussion of the cricket player.

Thus far then, Deleuze's position seems reasonably congruent with Merleau-Ponty's and Dreyfus' analyses of embodied learning, but there is at least one obvious and important difference between their respective analyses that needs to be considered. Deleuze focuses on apprenticeship rather than mastery, and frequently talks of what he calls an "essential apprenticeship" (DR 164). On his view, the apprentice is someone who is not preoccupied with knowledge, which is said to create generalities in the form of rule-following or rule-enabling solutions, but is instead envisaged as someone who occupies and inhabits problems in a more practical and experiential way. Although this might not appear to be antithetical to Dreyfus' position, Deleuze's aversion to thematizing mastery and expertise cannot be said to rest on merely a rhetorical difference with Dreyfus. Consider Deleuze's preferred example of learning, which is about learning to swim. When a body swims with a wave it comes up against that which is radically other than it and it also involves difference in the repetition from one wave to the next. He returns to this example toward the end of *Difference and Repetition*, where he discusses the way learning evolves entirely in the comprehension of problems, and he suggests that learning to swim, or learning a foreign language, means "composing the singular points of one's own body or one's own language with those of another shape or element, which tears us apart but also propels us into a hitherto unknown and unheard-of world of problems" (DR 192). While on one level it is hard to argue with this analysis, the kind of learning experience with which Deleuze is primarily concerned is akin to that of the beginner (at least according to the Dreyfusian schema), but for Deleuze himself it also seems clear that this so-called beginning is also the end, paradoxically enough, and that which constitutes true expertise and stands as an exemplar for a life of encounters and exposure to difference. While Patton may be right to insist that a philosophy of difference need not entail a politics or a morality of difference,[23] it is difficult, if not impossible, to keep them separate in various passages of Deleuze's work. This coimplication of a tacit morality and transcendental philosophy of difference is particularly in evidence in his repeated insistence that the genuinely important aspect of learning is the way in which it always involves a violent training, a traumatic experiment, and an apprenticeship that "gives rise to images of death" (DR 23). Although this last comment invokes the trauma that is also at the heart of Heidegger's famous theorization of being-towards-death in *Being and Time*, there are nevertheless some important differences between their positions. After all, at one stage Deleuze compares the experience of learning to the acephalic, the albino, the aphasic (DR 165), and this understanding of learning rejects Dreyfus' (and to a lesser extent, Heidegger's) more normative claim that a suitably refined adjustment toward one's environment is the telos of learning and skill acquisition. On Dreyfus' view, learning and skill development may be motivated by problems, *Angst*, or gestalt instabilities, but it seeks to defuse them of their incendiary power and this is precisely what Deleuze seems to want to deny and downplay. And while there are some good reasons behind Deleuze's aversion to the language of mastery and expertise (politically it carries a certain distasteful resonance of elitism, but we must see that for Dreyfus such comportment is also foundational to each and all of our experiences), it seems that Deleuze's attempts to radically distinguish certain learning experiences from habit and coping is based on an inadequate understanding of coping as purely about resemblance and equivalence, and hence as instigating a realm of generality. Both phenomenological reflection and Dreyfus' empirical studies seem to suggest that this need not be the case.

In-itself, this may not be all that significant. However, Deleuze's conviction that the experience of learning must be understood on the model of the trauma of apprenticeship (an apprenticeship that resists co-option to the living-present of *l'habitude* and skilfull coping) results in a consequent, and I think philosophically problematic, valorization of the child-like and something akin to a tabula rasa state that is not subject to the sedimentation of history and the weight of inhabiting. Deleuze tells us, for example, that the child-player can only win, and he exalts both the dice-throw and jumping into the volcano, both of which evince the true spontaneity of the child (DR 116). His conviction is that when chance is affirmed, one cannot lose. It is hard not to wonder whether such a position constitutes a retreat into romanticism, or, better, empirico-romanticism. Of course, Deleuze's philosophical intention with this material on the dice-throw and the child is to link it up with his general privileging of the time of the future, and this is also the point of his reconfiguration of Nietzsche's thought of the eternal return of the same along the lines of the return of difference. While Deleuze's transcendental philosophy of time is one of the more prescient philosophical contributions of the last generation or two, it seems to me that although he strongly critiques representationalist and rule-governed understandings of action and behavior, there is a sense in which he nevertheless remains within this paradigm when he tries to simply reverse it by exalting the trauma of apprenticeship, the value of the child and the child-like, and privileging singular encounters and experiments (this becomes more apparent in his collaborative writings with Guattari, although I cannot pursue that here).[24] While his particular version of empiricism explicitly distances itself from the blank-slate picture of the subject of traditional empiricism, one can discern its return in his moral dedication to nomadic living, which subtly, but persistently, evinces a nostalgic desire for getting as near as possible to such a state, and for an ethic of maximizing "forgetting and connecting,"[25] something that seems closely connected to the polymorphous perversity that Freud attributes to infants. Forgetting and connecting are certainly important, perhaps even worth maximizing, but we cannot ignore this bodily intentionality that precludes random connections.

Of course, it may be protested that I am missing the main point of Deleuze's work, which is that what is important for him is the ability of a being to go beyond or exceed its limits. Certainly this is the Spinozian and Nietzschean affirmation at work in his *Nietzsche and Philosophy,* and it is also in this respect that Deleuze poses the following important rhetorical question, "to what are we dedicated if not to those problems which demand the very transformation of our body and our language?" (DR 192). It is an important question, linking in as it does with the Spinozian insistence that we may not yet know what a body can do, but, and importantly, Deleuze assumes the answer to this question—and assumes its ethical priority—without ever justifying it. Why are we above all dedicated to that which disrupts and exceeds limits? Why not focus on the way in which beings and bodies, rather than exceed or transgress their limits, constantly extend, modify, and redefine their limits (as we are doing all of the time with our neural networks)[26] in a manner that is well-captured and described by Dreyfus and Merleau-Ponty's model of experience providing increasingly refined embodied solicitations to act? Change is not only possible but inevitable on this understanding, and the advantage of this latter perspective is that it doesn't entail the simultaneous moral and transcendental positing of some kind of radically disruptive event, akin to a first genetic cause, that forces us to acquire new faculties and new modes of being.

Moreover, even on Deleuze's own analysis something like the privilege that most phenomenologists accord to embodied coping (and to a movement toward equilibrium) finds its way back into the picture. After all, we have seen that in *Nietzsche and Philosophy* he endorses the Nietzschean conviction that a body will increase in power only to the extent that its capacities to affect and be affected become more developed and differentiated. We have also cited Patton's remark that "there is a feedback loop between the success or otherwise of attempts to act and the agent's capacity to act." There are clear links between this conception and the feedback loop that Merleau-Ponty and Dreyfus consistently refer to in their understanding of the body's ongoing maintenance of intentional arcs. This raises at least a couple of important questions: first, and more empirically, are we ever really disrupted in the radical way that Deleuze privileges, even in the experience of learning, or is it not that embodied skills and intentionality are always already mobilized and at work; and, second, why think that this fluid mobilization of capacities is somehow insufficient? Isn't this, surprisingly enough, itself a form of slave morality? A way of reposing this question in order to make explicit its significance for ethics and politics, is to ask why the move to establish an equilibrium is of less value than the poststructuralist insistence on that which disrupts the equilibrium? It is a difficult question to answer, but it seems to me that any philosophical position that a priori excludes this kind of gathering (the importance and priority of *l'habitude* and of coping) verges on being an anarchism when it comes to the political realm, regardless of their explicit position on the role of the state.[27] Of course, that does not make any such position wrong or politically ineffective. Both anarchist and poststructuralist problematizations are politically vital, even if they have more of a moral than political force. Nonetheless it seems to me that any adequate political philosophy could also do with a phenomenology immured of its communitarian spirit through due recognition of the inevitability and vitality of social conflict, but one that, against Deleuze, recognizes the inextricability of habitual coping and learning, and also comprehends the contribution of the former to *both* inertia (as the poststructuralists have shown us, perhaps especially Foucault)[28] and productive change (embodied coping is not necessarily purely reactive, a form of slave morality). Such a perspective ensures that philosophical and social reflection are not so abstracted from history as to suggest the model of pure reason, but in thematizing embodied coping properly one also avoids the somewhat romantic morality that seems to follow, at least in some of Deleuze's writings, from the overturning of that tradition.

11

Touched by Time: Some Critical Reflections on Derrida's Engagement with Merleau-Ponty in *Le Toucher*[1]

The philosophical relationship that obtains between the work of Merleau-Ponty and Derrida has continued to intrigue and preoccupy many of us despite, or perhaps even partly because of, the fact that Derrida did not accord the work of Merleau-Ponty much attention during his remarkably prolific career. Two relatively recent books of Derrida's have addressed this gap: *Memoirs of the Blind* and, more recently, *On Touching*. However, although Derrida proposes an "entire re-reading" of the later Merleau-Ponty in *Memoirs of the Blind*,[2] with the clear implication that there are hitherto unaccessed and invaluable resources to be mined in this body of work, I will suggest that the actual reading of Merleau-Ponty propounded in *On Touching* falls well short of this ambition. While this chapter will raise some critical questions about the interpretation that Derrida offers of Merleau-Ponty in "Exemplary Stories of the Flesh: Tangent 3," including the implication that his work on the senses and intersubjectivity remains mired in theological prejudices, it will also be concerned to examine the transcendental philosophy of time (or philosophy of the *contretemps* that breaks open time but nonetheless pertains to it) that undergirds and motivates Derrida's engagement with the philosophies of touch. In this latter respect, I will argue that Derrida's philosophy is itself touched by time, in the peculiar sense of "touched" that connotes affected and wounded. On my reading, his work instantiates an ethics of nonpresentist time,[3] an ethics of that time which is the transcendental condition of the present and any event of touch. I ask whether this prevarication on the issue of the transcendental and the ethical is reason to look for a different understanding of both time and the transcendental to Derrida's, and I end this chapter by once more proposing a dialectic between the disjunctive and conjunctive aspects of time (wound and scar) that does not accord any kind of a priori privilege to the one over the other.

Touching the Untouchable: Phenomenology as Haptology?

First, however, let me briefly try to summarize *On Touching*. Aside from the detailed engagement with the work of Jean-Luc Nancy that will not be our prime concern, *On Touching* also offers a remarkable genealogy of the links between an optical intuitionism that has arguably undergirded large parts of modern Western philosophy and a less regularly observed haptocentrism that simultaneously grants a privilege to touch. In other words, Derrida argues that the "exorbitant privilege" accorded to sight and optics is often also (perhaps inevitably) supplemented by a desire for presence (the intentional movement of desire) in which touching and the human hand play a foundational role. Whether he is analyzing the work of Plato, Descartes, de Biran, Kant, Diderot, Berkeley, Bergson, Husserl, Heidegger, Merleau-Ponty, Chrétien, Frank, and others, according to Derrida there is never "any privilege for the gaze (no optical theoretism) without an invincible intuitionism that is accomplished, fulfilled, fully effectuated, starting from a haptical origin or telos" (OT 204). To put it another way, Derrida argues that to varying degrees each of these thinkers remains a Christian ontotheologist, yearning for some kind of coincidence and a self-touching communion. The motif of the flesh employed by Merleau-Ponty, Chrétien, and others, is, for him (as for Deleuze and Guattari),[4] particularly symptomatic of this.

It is important to note at the outset that, for Derrida, it is time that ultimately precludes and undermines this desire for coincidence, even though he only rarely directly states this in *On Touching*. Consider his comment: "the principle of intuition, finds itself threatened—as it happens to be, *once again*, by the experience of temporalisation that is indissociable from this" (OT 192). The "once again" here evokes his earlier engagement with Husserl in *Speech and Phenomena* and the *différance* that interrupted the possibility of an exact internal adequation with oneself that Husserl's phenomenology sought to secure. Indeed, Derrida's recourse to a philosophy of time in order to undermine intuition (that phenomenological principle of principles) also parallels his claim in the same book that, "in the last analysis, what is at stake is . . . the privilege of the actual present, the now."[5] In fact, we have suggested that all of Derrida's most telling and repeatedly expressed objections about phenomenology are about time, and this background is necessary in order to understand what is at stake in Derrida's critical engagement with Merleau-Ponty. After all, without doing Derrida's analyses too great an injustice, his position can be schematically represented as follows: all philosophies of touch (and of the body) are, to greater and lesser extents, intuitionisms. They presuppose some kind of timelessness or immediacy in a given encounter/

touch (self-presence), or they idealize such a situation as the infinitely deferred horizon toward which we can and should aspire. We can also generalize and say that on Derrida's understanding all phenomenology, whatever its unarguable merits, amounts to a haptology, a philosophy of touch that privileges the hand (and therefore an intuitionism). As would be apparent from the conjunction of these two claims, Derrida likewise contends that Merleau-Ponty's work is irremediably marked by both of these tendencies. Indeed, on Derrida's analysis, at least as it is presented here, Merleau-Ponty's phenomenology remains fundamentally committed to forms of presentism and to coincidence rather than noncoincidence (OT 211). In other words, the allegation is that his work harbors an omnipresent theological desire for union, something that for Derrida betrays itself in the later Merleau-Ponty's deployment of the Christian notion of flesh.

As a consequence of this guiding interpretative ambit, there are a few criticisms of Merleau-Ponty in *On Touching* that are rather ungenerous and which it is worth briefly recounting. Notwithstanding his recognition that there are aspects of Merleau-Ponty's work that pull in contrary directions, Derrida's basic position is that Merleau-Ponty's philosophy of ambiguity (and confusion) entails a valorization of self-presence (OT 193, 198). Now, such a position omits to consider the constitutive role that time plays in Merleau-Ponty's own arguments for ambiguity in *Phenomenology of Perception* (PP 345, 410-33). Moreover, when Derrida rails against Merleau-Ponty's doctrine of the confusion of the senses (OT 198), exemplified by his declaration that synaesthesia is the rule (PP 229), Derrida does not consider any of the surfeit of phenomenological and scientific evidence that, both before and since Merleau-Ponty's *Phenomenology of Perception*, has provided support for such a position. Developmental psychology and cognitive analyses of neonatal life continue to consistently point to an originary confusion and synaesthesia.[6] And it seems to me that phenomenological philosophy ought to be interested and concerned with empirical data of this kind. If the grand phenomenological ambitions of Heidegger and Husserl, and the very clear distinction between the transcendental and the empirical is rather more complicated than their work suggests, then any existential phenomenology that is oriented around the body can and should consider other resources: psychoanalysis, neurology, cognitive science, analytic philosophy of mind, and so on.

Of course, whether the emerging consensus around synaesthesia constitutes an objection to Derrida depends upon his understanding of the nature of transcendental argumentation in general (and the manner in which Derrida himself deploys such arguments in relation to a privilege he grants to that which breaks open time, to *différance*), since it seems tantamount to protesting that Derrida's analyses of the transcendental are not empirical enough.

Derrida might respond to such a reproach that there can be no *concept* of separate senses if they are originally confused in the manner Merleau-Ponty suggests. He might hence argue, modestly enough, that in order to be consistently thought, the senses must be distinguishable in principle, although this need say nothing about how the senses are experienced in fact, or even the conditions for such actual experience.[7] There is, after all, an important difference between what the literature refers to as concept-directed and truth-directed transcendental arguments.[8] If Derrida confines his analyses to the more modest ambitions of the former, without any deeper metaphysical/ontological connotations about the ways things really are, or the conditions of possibility for such reality/presence, then there is not necessarily a contradiction here between the findings of the empirical sciences and his transcendental analyses; on the contrary, it would make the former almost beside the point. Now that may be an unwanted consequence in itself, and indeed it is one of the worries that analytic philosophers have long held about phenomenology and continental philosophy more generally,[9] but I will show that Derrida deploys his transcendental arguments in both of these senses (genealogical/concept-directed and quasi-ontological). Because of this, it seems to me that he does have a need to grapple with the empirical findings of the sciences to a greater extent than he does. While to question Derrida from such a perspective is not to invalidate his work, it does suggest that there are a conglomerate of reasons that may favor another approach, and it is also important to the issue of his relation to Merleau-Ponty because most of Derrida's key disagreements with him tacitly depend on a disjunction between the transcendental and the empirical. As such, despite appearances, "Tangent 3" actually offers a curious lack of actual engagement between these two philosophies, not least because the significance of the transcendental to their work (and the relation of the transcendental to the empirical findings of the sciences) lurks in the background as arguably the key point of dispute but it is never foregrounded as such.

But to return to Derrida's problems with Merleau-Ponty, the latter is said to problematically put sight and touch on same plane (OT 186); there is hence a parallelism in Merleau-Ponty's work (OT 200). Now any such parallelism is not logically entailed by the doctrine of synaethesia (a blurring of perceptual boundaries does not entail that we cannot meaningfully speak of sight and touch), and nor is it clear that Merleau-Ponty puts them on the same plane, despite some of the awkward apparent postulations of equivalence in *Signs* that we will return to because they rightly catch Derrida's attention. But Derrida also claims that Merleau-Ponty relies upon a hierarchy of the senses (OT 206) and he says this specifically of Merleau-Ponty's unfinished but remarkably rich, *The Visible and the Invisible*. Derrida suggests that for Merleau-Ponty the first example remains the visible, but also that it is the hand that subsequently comes to dominate the discourse (grasping, pointing, touching: always of the human hand). But despite the title and the enduring motif of the two hands touching one another in Merleau-Ponty's work, it is not as clear as Derrida thinks that any such "handthropocentrism" or "human-nualism" persists in *The Visible and the Invisible* (which we should note was originally titled "The Origin of Truth"). What about the role of music and sound, both of which are vital to Merleau-Ponty's famous final chapter, "The Intertwining: The Chiasm"? Does the example of the (human) hand really drive the discourse for the entirety of this chapter as Derrida suggests? (OT 208). Is this humanism itself an implicit theological remainder in Merleau-Ponty's work? It seems to me that this must remain an open question, rather than something that is "hardly disputable" as Derrida declares at one point (OT 203).

Of course, this does not vindicate Merleau-Ponty and offering a textual defence of Merleau-Ponty against Derrida's interpretation is not the main purpose of this chapter in any case. It is more fruitful to acknowledge that these criticisms do touch on some substantive philosophical differences between Derrida and his predecessor that much of the remainder of this chapter will be preoccupied with. These include: the allegation that Merleau-Ponty reduces the phenomena of touching to the hand and, therefore (on Derrida's account), to the mastery of the "I can"; the suggestion that Merleau-Ponty's embodied phenomenology privileges gathering and coincidence rather than noncoincidence and dispersal, and his argument that this becomes most apparent in what he considers to be Merleau-Ponty's "intuitionism" in regard to the other person. Both of these key criticisms relate to time. For example, in granting a primacy to pragmatic skills and equilibria within an environment (the "I can" or motor intentionality of prereflective consciousness), for Derrida Merleau-Ponty's philosophy presents a simplified account of touch not merely because it prioritizes acts like grasping, but ultimately because it constitutes a philosophy of touch that brackets away the desire to touch and the injunction not to touch in favor of adjusting toward one's environment. That is, it will be a philosophy of touch that is bereft of time, in that it will have been divested of the traces and residues of the past (that which tradition has forbidden us to do and which is thus promised above all else), along with the futural touch (the prospect of an event) which haunts and interrupts the stable order of time. Rather than the telos of touching being equilibria as *Phenomenology of Perception* at least seems to suggest, for Derrida it might be more accurate to say that it is the disequilibria that is bound up with the paradoxical desire to touch the untouchable (as with Levinas in *Totality and Infinity*, desire is oriented around an exploration of the unknown).

Now, at first glance Derrida's implication that Merleau-Ponty's account of touch is one without time seems somewhat disingenuous, especially given

Merleau-Ponty's conception of the phenomenological reduction (history and social life cannot be bracketed away for a pure phenomenological analysis of touch), as well as his insistence on ambiguity in his early work, and reversibility in his later work. Both positions put time at the center of sensibility and ensure that there is no touching that is unambiguously self-present. They form one part of his ontological argument that the experience of being touched, either previously or anticipated, encroaches and overlaps with the experience of touching—as does culture, even if they are based on proprioceptive capacities of bodies, such as the body-schema, that are, to some extent, precultural, in that they are already apparent from the earliest stages of foetal life. Nonetheless, some important questions remain about the "I can" of embodied life and the prereflective motor intentionality that is, at least to an extent, prioritized in *Phenomenology of Perception* (it is not so clearly the case in *The Visible and the Invisible*). Derrida's basic problem is that this priority assumes a stability to the senses that has no role for the untouchable and the impossible. To put it another way, we might say that, for Derrida, touching is far too complicated to account for under the schematics of the prereflective body preparing to grasp or point to an object—and, we might note in passing, the phenomenological distinction between grasping and pointing has engendered much Merleau-Ponty-inspired research in the cognitive sciences and other disciplines.[10] On Derrida's characterization, sense-certainty is precisely that for Merleau-Ponty—certain—rather than penetrated or wounded by the desire to touch and the injunction not to touch. Derrida's fundamental argument in *On Touching* is that there is no touch without the wound of time (dispersal, *différance*). This is another way of saying that there is no touch without the impossible and the untouchable, noting that the impossible term of the aporia is consistently associated with the past and the future. For Derrida, the past and the future, desire and repression, saturate the senses in such a manner that they are not assured and confident, but, more likely, stuttering and tentative, in a manner that is closely related to Deleuze's valorization of the trauma of apprenticeship in *Difference and Repetition*. The senses are affected such that they are in disarray rather than in relatively harmonious interaction and functioning as an integrated whole, as is presupposed by much of Merleau-Ponty's early phenomenology, arguably even when he is considering abnormal cases like Schneider and the experience of phantom limbs in order to shed light on what is presupposed when the body is inconspicuous or "transparent" to us.

There are, however, some things that perhaps ought to be said in Merleau-Ponty's defense here. Notably, Derrida gives no attention to what we might call the sexual-schema, which functions by transposition (PP 168) and substitution (PP 78), unlike the body-schema that functions by proprioception (crudely, our prereflective sense of our body). Our proprioceptive capacities and their movement toward establishing an equilibrium within a given environment are never self-sufficient; rather they are always allied with and supplemented by the sexual schema, including desire for others, the unknown, and so on (PP 165). This is why Merleau-Ponty says existence is suffused with sexuality (PP 159) but nonetheless is not reducible to it (PP 166, 169). Arguably, this divests Merleau-Ponty's conception of the body of the assumption of "properness," which Derrida is so worried about. We will return to this point in the context of the purported lack of introjection in Merleau-Ponty's philosophy, something that also greatly concerns Derrida, but suffice it to say that in sexuality (understood in the broadest sense) the body is dispossessed, in that desire is experienced through a series of substitutions.[11]

Nonetheless, let us grant to Derrida that such aspects of Merleau-Ponty's work are not given as much attention as the discussions of the body-schema and our prereflective attempts to secure equilibria through the establishment of habits, maximum grip, intentional arcs, skills and the like. Confronted with this disparity, it is a difficult question as to how to decide between these perspectives. Surely Merleau-Ponty's rich analyses of the body compellingly show us our flexibility and adaptability. Likewise, it is difficult to deny that phenomenologically there is an integrity and unity to perception that empiricism has been unable to adequately countenance, and that intellectualism has falsified in giving the preeminent constituting role in this unity to reflective judgment, as in Merleau-Ponty's famous and influential critique. There is also an important phenomenological distinction between pointing and grasping, and between the reflective and the prereflective, that it seems that Derrida wants to problematize. But one would not want to simply give these up without some good reasons and/or fairly compelling counter-evidence. On what basis then, can Derrida overturn or subtly displace the force of this analysis?

Derrida's point here as elsewhere is that phenomenology covers things over. Something resists or elides phenomenological analysis, notably a certain trace and a certain traumatism. What kind of argument is this? It is a transcendental one, grounded in all of his prior attempts to show that *différance* is the condition of possibility of presence, and also renders that presence unstable and liable to transformation—hence his early use of the phrase "quasi-transcendental." To be more accurate, it is simultaneously a transcendental and a genealogical argument that Derrida provides us with (again, quasi-transcendental). As Derrida shows us in *On Touching*, when we speak of tact, for example, it is a matter of knowing how to touch on the untouchable, which cannot be a matter of simply touching it (OT 67). Can a phenomenology of perception tell us anything about tact, Derrida's rhetorical question seems to be? And without this, can a phenomenology of touch really have anything to say about the concept of touch, perhaps even the experience of touch? What if tact is involved in all touching, such that the sense of touch is problematized by the untouchable, the cultural and symbolic? These are, I think, important questions. In fact, Derrida is surely right in this regard, but it is not clear that Merleau-Ponty's philosophy (even his earlier phenomenology) is committed to disputing this. For him, the phenomenological reduction functions by showing us its impossibility, by showing us our ties to the world, history and time (PP xiv). Moreover, the embodied attempt to secure an equilibrium—we adjust our gaze to observe something appropriately, to get maximum grip—need not be committed to reducing all touch to this kind of teleological activity. On the contrary, we have seen that the sexual schema functions in a very different way. The point is just that it is a basic modality of being-in-the-world, and the onus is on Derrida, I think, to provide an argument as to why it should be relegated to secondary status, replaced by a philosophy of the prosthetic rather than a philosophy of the body (OT 237). As an aside, we might note that the prosthetic can also be seen as central to Merleau-Ponty's analyses in the exploration of phantom limbs and the manner in which cars, glasses, and the like become part of an expanded body-schema. Despite Derrida's strong criticisms then, it seems that there are many aspects of Merleau-Ponty's work that elide his analysis and indicate that it is not clear that all philosophies of the body must be premised upon a haptical intuitionism and implicitly theological, yearning for some form of communion with nature, world, or God. To put it another way, the fact that philosophies of the body (phenomenological and otherwise) have often been problematic in the ways that Derrida indicates does not show that their considerations are either philosophically unimportant or that they should instead be consigned to the lesser term of a dualism (e.g. as a present, empirical state of affairs, in contrast to the transcendental condition, the *contretemps* that breaks open the present toward the past and future).

Moreover, some further questions about the Derridean strategy remain to be asked. For Derrida, these intriguing discussions surrounding tact and the paradoxical desire to touch the untouchable are again largely motivated by his concern for the event and for the new. Consider *On Touching*, where Derrida cites Nancy's declaration that "the impossible is what takes place." Derrida responds: "Madness. I am tempted to say of this utterance, itself impossible, that it touches on the very condition of thinking the event. There where the possible is all that happens, nothing happens, nothing that is not the impoverished unfurling of the predictable predicate of what finds itself already there, potentially, and thus produces nothing new" (OT 57). Derrida's language here is strong: where the possible is all that happens nothing happens, except the "impoverished unfurling of the predictable predicate." There is the event, there is the new: these are of the order of the impossible (conceptually speaking), but they are also what can and must motivate us (experientially speaking) as Derrida insists in so many different places in his later work. What is the status of this kind of transcendental argument? It is not clear, being simultaneously conceptual, metaphysical/ontological (e.g., the disruptive time of the event is a condition of the possibility of presence) and ethical. Is this opposition between the event and the predictable predicate a necessary one, either conceptually or for the possibility of experience per se? Derrida says it is a condition of thinking the event, but is this really the only manner in which one can account for, or think, the event, and what gives this manner of thinking the event a privileged ethical role? Is every philosophy of mediation, of continuums, condemned to be unable to explain the event? It is not at all clear that this is so. Presumably tipping points can still be theorized. Moreover, if the event can only be thought on the basis of such an abstraction (the event versus the predictable predicate), as Derrida suggests, then such a thought would, having little to do with the experiential and not rigorously transcendentally grounded, arguably be metaphysical.

This does not, of course, prove Derrida wrong, but is it not the case that the body (including the body-schema) is itself a condition of possibility of both experience and the enunciation of such transcendental arguments? In such a formulation the boundaries between the empirical and the transcendental are clearly blurred, but it is curious that Derrida does not consider this possible path. On the contrary his position (event versus predicate) seems to inexorably come from Levinas' rejection of the "imperialism of the same" and is, arguably, a defining feature of poststructuralism. For Derrida, for there to be an event of touch it must, paradoxically, involve the impossible: touching the untouchable. As I have noted, there is a conceptual logic to this position that is compelling, but what remains to be explained is the priority that the untouchable and the impossible continue to enjoy in Derrida's work in both an ethical and a transcendental sense (beyond that of being a mere necessary condition for conceptual thought). And why is it, by contrast, that the recuperative capacities of bodies (habits, skills, etc.) and their "omni-temporal time" as Derrida refers to them in *Politics of Friendship* remain devalued, with virtually no attention paid to them? (PF 16, 20). This question is never really answered in Derrida's work, although it is tacitly assumed that there are ethical considerations in favor of his position. Let us broaden our inquiry then, in order to see the ethical and intersubjective significance that might be accorded to Derrida's transcendental philosophy of time.

Touching the Untouchable Other

After all, bound up with his aporetic analyses of touching the untouchable, is also the problem of the other (whether person or nonperson). For Derrida, the problems of time and the other are interrelated,[12] and it is hence unsurprising that he also takes Merleau-Ponty to task for his conception of intersubjectivity along essentially Levinasian lines.[13] For Derrida, Merleau-Ponty's view too often domesticates the other's essential untouchability, making of them an analogue of our own flesh and positing a communion without divergence (*écart*): Merleau-Ponty "*runs the risk* of reconstituting an intuitionism of immediate access to the other . . . *runs the risk* of reappropriating the alterity of the other more surely, more blindly, or even more violently than ever" (OT 191, my emphasis). This is curious language for Derrida, this repeated "runs the risk" that is also in the original French *Le Toucher*.[14] This is not as careful as he usually is, and its lack of rigor perhaps testifies to the uncharitability of his reading of Merleau-Ponty. It conveniently ignores that Merleau-Ponty explicitly targets all philosophies of intuition in *The Visible and the Invisible* in a chapter entitled "Interrogation and Intuition." Perhaps even more significantly, Derrida never mentions two key interrelated ideas that preoccupy the entirety of this book, chiasm and reversibility, an enumeration of which would surely threaten to undermine his argument, given the centrality of *écart* to the reversibility thesis and the genesis of sense.[15] In fact, the textual support that Derrida brings forth to justify his interpretation is based almost exclusively on the essay, "The Philosopher and His Shadow" (from *Signs*). Although this makes sense given Derrida's attempts to show how Merleau-Ponty has misread Husserl (and in this respect Derrida is surely right), it is curious that he devotes such cursory consideration to the ontology of *The Visible and the Invisible*.[16]

Fundamentally, Derrida contests Merleau-Ponty's account of relations with others because of the way that Merleau-Ponty says that we have them immediately and without introjection. Derrida cites and italicizes Merleau-Ponty's comment about the phenomenological apprehension of the other: "It is imperative to recognize that we have here neither comparison nor analogy, nor projection nor 'introjection.'" For Derrida, this entails "an intuitionism of immediate access to the other" (OT 191) and it is this immediate phenomenological access to the other person "without introjection" that Derrida is worried about. As Derrida suggests:

> If there is some introjection and thus some analogical appresentation starting at the threshold of the touching-touched, then the touching-touched cannot be accessible for an originary, immediate, and full intuition, any more than the alter-ego. We are here within the zone of the immense problem of phenomenological inter-subjectivity (of the other and of time): . . . shouldn't a certain introjective empathy, a certain "inter-subjectivity," already have introduced an other and an analogical appresentation into the touching-touched for the touching-touched to give rise to an experience of the body proper allowing one to say, "it is I," "this is my body?" (OT 176-7).

Derrida suggests here that there is some analogical appresentation at the heart of the touching-touched, and at the heart of the seeing another, or touching another. The technical, the prosthetic, is installed at the heart of the purported pure intuition of bodily apprehension, something akin to what Derrida elsewhere refers to as the "prosthesis of origin,"[17] but let us think a little further about what this might involve, why we should believe it, and its relationship to the positions that Merleau-Ponty holds. How, for example, does an "introjective empathy" become an "analogical appresentation"? Are these equivalent as Derrida's comments seem to suggest? It seems to me that the former need not entail the latter, and that certain argumentative premises are missing here.

We should also note that when Merleau-Ponty says "without 'introjection'" in the passages that so trouble Derrida, the term introjection is persistently placed inside quotation marks suggesting that it has a particular meaning that is not obvious. Indeed, when Merleau-Ponty uses the terms shortly afterward, it is in the context that projection and introjection are understood representationally and to be opposed to the prereflective. Merleau-Ponty says, "I can think that he thinks; I can construct behind this mannequin a presence to self modelled on my own; but it is still myself that I put in it, and it is then that there really is 'introjection'" (*Signs* 169). His real target here hence seems not to be introjection loosely understood, but arguments by analogy that depend upon some kind of representation of the other and that are premised on reflection and abstraction.

For him, of course, there is an overlapping between self and other, a transitivism between self and other, which must occur developmentally and even in adult conversation with another. We continue to "borrow ourselves from others" and to "create others from our own thoughts" (*Signs* 159), and his philosophy is hence premised upon a minimal introjection and projection, if we understand those terms to refer merely to the manner in which self and other are relationally constituted, both developmentally and phenomenologically. Merleau-Ponty consistently draws attention to the way in which newborn babies are able to imitate the facial expression of others, both in *Phenomenology of Perception* and in his essay, "The Child's Relations with Others," something that provides the basis for our relations to others thereafter. As Thomas Fuchs notes in this context, "by the mimetic capacity of their body, they transpose the seen gestures and mimics of others into their own proprioception and movement."[18] There is much emerging evidence for this ability to put ourselves virtually in the place of another,[19] particularly in association with recently discovered mirror neurons that "fire" (in many cases) in the same manner when we perceive certain kinds of action as when we perform them ourselves. As Fuchs suggests, there hence seems to be a virtual modeling of our motor-schema in response to the other's motor-schema and expressions. In his words, "the body works as a tacitly 'felt mirror' of the other. It elicits a non-inferential process of empathic perception" in this transfer of corporeal schema. On one level, it is hence difficult to see what the force of Derrida's objection might be. On Merleau-Ponty's account, the other is there from the "beginning" (as Derrida would want), whether through the sexual schema that is copresent with the body-schema, or through this virtual modeling of the other that is nonetheless immediate and noninferential.

Merleau-Ponty's concern with the narrower and more technical understanding of "introjection," however, seems to be with the way in which it connotes a distinct inner and outer, where a subject replicates in itself behaviors of others in the surrounding world (this is perhaps the analogical appresentation that Derrida invokes, but might this not be said to be incorporation rather than imitative and prereflective introjection?)[20] This sometimes happens, of course, but Merleau-Ponty's point in this passage is that there must be a prior overlapping and encroachment (a chiasm, not a union) for this to be possible, exemplified in this instance by the non-inferential relations between self and other that obtain from birth, and there is an ongoing exchange between self and other. For Merleau-Ponty, this is both a transcendental argument, and it is also simultaneously said to be phenomenologically given (on his analysis) and susceptible to scientific analyses that give the argument more or less weight. This becomes clearer in the second half of the passage that Derrida discusses, where Merleau-Ponty goes on to say: "On the other hand, I know unquestionably that that man over there sees, that my sensible world is also his, because I am present at his seeing, it is visible in the eyes' grasp of the scene. And when I say that he sees, there is no longer here (as there is in 'I think that he thinks') the interlocking of two propositions but the mutual unfocusing of a 'main' and a 'subordinate' viewing" (*Signs* 196). The other person exists for us before the order of thought, but that is not necessarily to say without the mediations of culture, which also has a prereflective weight.

What might Derrida contest in this? Is Derrida suggesting that there can be no distinction between motor intentionality (or the so-called prereflective) and reflective analogical thinking at all, between introjection and incorporation (analogical appresentation)? Or is he just calling into question the ability of a reflective activity like philosophy to describe this prereflective aspect without falsifying it? If he subscribes to the latter, then we should note that for Merleau-Ponty the perceptual faith is described precisely as a paradox, and it is a paradox that he gives a lot of attention to via his notion of the hyper-dialectic in *The Visible and the Invisible*—the key point, however, is that the philosopher cannot simply return to this faith. Derrida is hence correct to point to the description of the prereflective as a difficulty for any philosophy of touch, but he does not address the manner in which Merleau-Ponty himself grapples with this question, and does not understand the prereflective faith as something that is simply given to us. On the other hand, it also sometimes seems that Derrida's insistence that there is analogical appresentation at the heart of all experience is an attempt to contest this distinction between the reflective and the pre-reflective. But this move seems implausible and there is little to support Derrida's intermittent declarations in this regard. In this text it is almost an axiom for Derrida, the consequence of various different transcendental arguments deployed throughout his career that are the background to this declaration. Nevertheless, it should not be ignored that it is a position that remains disjunct from experience and bodies, from phenomenology, developmental psychology, and cognitive science, among other disciplines. Derrida might accept this as something akin to a badge of honor, but from a Merleau-Pontyian perspective it means that Derrida remains an intellectualist who privileges reflection, what Merleau-Ponty calls a "high-altitude" thinker. Perhaps the theoretical price for accepting Derrida's view are too high to pay, at least without tinkering at the edges and making deconstruction a little more pragmatic by insisting on the import and value of lived time but without reifying it. That, at least, is my tacit proposal here: a transcendental pragmatics of an embodied variety, rather than either the Derridean "quasi-transcendental" or the Habermasian discourse ethics variety.

Now none of this is to suggest that Derrida doesn't correctly point out problems with several of Merleau-Ponty's formulations regarding the other, but the point remains that they are taken out of context because the major arguments being propounded by these formulations—e.g., chiasm/reversibility, the reflective and prereflective distinction—are not explicitly considered. For example, when Derrida (like Levinas) seizes on Merleau-Ponty's claim that, "I see that this man over there sees, *as* I touch my left hand while it is touching my right" (OT 197), his problem here is with the "as" that suggests the relation between my left hand touching my right hand is the same as that which obtains when I shake another's hand (or see another person, etc). These apparent declarations of equivalence are problematic, but focusing on them also misses Merleau-Ponty's key point. Both Merleau-Ponty's "as" and his "in no different fashion" (OT 190) denote a structural isomorphism that obtains (the chiasm, reversibility) phenomenologically, and apparently neurobiologically,[21] but this does not entail that every chiasm and every reversibility is equivalent. Merleau-Ponty repeatedly shows this is not so. The gap between touching and being touched in our own perceptual experiences of our body is not the same as that which obtains with touching or seeing others (there is a phenomenological difference that nobody could deny), but Merleau-Ponty's claim is that the former gap (*écart*) and intertwining (these come together for him), does "prepare" us for the understanding that there are others.[22] Merleau-Ponty explicitly says this in "The Philosopher and His Shadow" (*Signs* 168), but Derrida does not mention it despite focusing on this passage. Preparation is not equivalence. While it is difficult to dispute that Merleau-Ponty's language sometimes betrays an impulse toward some kind of intuitionism of the other (and, as Derrida points out, it is also significant that religious images of "communion," "original ecstasy," and "natal bond" abound), this is clearly an unsympathetic reading of Merleau-Ponty being propounded here. The question then is why. While we cannot, of course, know precisely what motivated Derrida, nor "what makes reading Merleau-Ponty so troublesome" for him, so "passionately exciting and difficult, yet also disappointing or irritating" (OT 211), we are now in a position to at least venture a tentative hypothesis based upon the tacit argumentative premises of Derrida's own text.

Rather than it being the case that Merleau-Ponty's work is insufficiently deconstructive and remains tethered to theological and humanist prejudices, it is more accurate to observe that Derrida's work is touched by a transcendental philosophy of time in a way that Merleau-Ponty's philosophy is not. Or, to put it another way, there is something akin to what Husserl might call a transcendental pathology in Derrida's philosophy, an antipresentism that eschews the time of Chronos, the orderly succession of presents in linear time, the habitual certitudes of the senses (including common sense). Despite their seeming incompatibility, Derrida accords both a transcendental and ethical privilege to a certain wound of time that he has given various names—*différance*, trace, etc.—and that recently has resulted in arguments problematizing embodied equilibria and any purported stability that we accord to the sense of touch and the senses more generally. It is difficult to dispute Derrida's forceful analyses of the coimbrication and contamination that time institutes. Nonetheless, because of the manner in which his transcendental arguments sometimes appear concept-directed, sometimes metaphysically or ontologically directed, a certain subterfuge takes places that renders both the findings of the sciences redundant, and that allows an ethical priority to be accorded to but one aspect of time, the disjunctive and the wounding (the past and the future: although they are not themselves present, they are the condition of presence). Derrida's arguments toward the ethical conclusion almost invariably beg the question. Just as Derrida accuses Merleau-Ponty of preferring coincidence to noncoincidence (presentism to nonpresentism) despite Merleau-Ponty's protestations to the contrary (OT 211), it seems to me that Derrida makes the reverse mistake. While Derrida's transcendental philosophy emphasizes the necessary intertwining of coincidence and noncoincidence more thoroughly than Nancy's does, it is nonetheless the case that the time of dispersal is ethically prioritized because of its links to the event (understood as rupture, as outside of the order of possibility). But the fact that the body excludes things from our particular horizons of significance with its habits and recuperative capacities that involve an omni-temporal binding that covers over gaps and aporias (without suggesting that they are not there) is not something that ought to be ignored. Likewise, the failure to consider the evidence that other disciplines help to provide for many of Merleau-Ponty's positions is problematic, despite Derrida's attempt to dress the decision up in transcendental terms. For me, it remains an unjustified leap of faith, and one that at least partly explains his enduring unease with the work of Merleau-Ponty.

12

Heidegger and Derrida on Being-towards-death and Philosophy's Untimely Future

Since Plato, Socrates, and perhaps before, the discipline of philosophy has frequently been associated with learning to come to terms with death. To some extent this soteriological tradition lives on in continental philosophy as practiced throughout much of the twentieth and twenty-first centuries, but this is much less the case in analytic philosophy despite the enduring interest in Epicurus' arguments for why we should not fear death, as well as in issues to do with metaphysics and personal identity that are raised by death. Indeed, following the French existentialist philosopher Gabriel Marcel's lecture on death, the British ordinary language philosopher, J. L. Austin, himself dying of cancer at the time, reputedly stood up and said, "we all know we have to die, but why do we have to sing songs about it?"[1] Without dwelling on Austin's attitude here, in this chapter I explore Derrida's engagement with Heidegger's treatment of being-towards-death in *Aporias*. In many respects it is the center of Heidegger's account of the future, the "not yet," as well as the key to his reflections on authenticity in regard to our mortal lives. Derrida's *Aporias* not only complicates the Platonic and Heideggerian conception in which philosophy is thought to help us to properly come to terms with death, but the discipline of philosophy itself is also revealed as aporetic in its essence, as not able to rigorously delimits its methods and concerns from those of other disciplines, whether they be naturalistic or socio-cultural in inspiration. In fact, for Derrida, both the phenomena of death and the discipline of philosophy are intrinsically aporetic. This does not mean the end or death of philosophy I hasten to add, but, if Derrida is correct (and I think he is in this respect) it does problematize, or at least complicate, some of the metaphilosophical distinctions that philosophers have typically used to define their terrain, including both Heidegger's guiding distinction between ontological and ontical inquiry in Heidegger, and the transcendental/empirical distinction in Kant that Heidegger also inherits and transforms.[2] Given the significance of transcendental reflection to the "divided house" that is contemporary philosophy it is fitting that we return to such questions in this, the penultimate chapter of the book. After all, despite Derrida's insights about what he calls a quasi-transcendental law of contamination between philosophy and non-philosophy, I will nonetheless suggest in conclusion that both large swathes of analytic philosophers and the poststructuralism of Deleuze and Derrida too readily lay claim to having attained to *the* philosophical stance, in contradistinction to each other as well as to phenomenology. Derrida criticizes Heidegger for precisely this gesture too, so in the course of this chapter it will be important to recognize that this book ultimately endorses a certain quasi-transcendental Derrida against other aspects of Derrida's transcendental philosophy of time that earlier chapters have expressed concerns about.

Two Aporias: Death and Philosophy

Derrida begins *Aporias* by considering the concept of an aporia that he takes from Aristotle and extends beyond the realm of logic where an aporia refers to a cognitive situation in which the threat of inconsistency confronts us, particularly one in which we have several individually plausible contentions that are collectively problematic.[3] Giving it something akin to an existential inflection, Derrida says an aporia is a puzzle, an impasse, a constitutive "not knowing where to go." (A 12) And this non-track, or barred path—*aporos* in Greek means "without passage"—is precisely what he is interested in in all of his work. In particular, it is his main concern when discussing "possible-impossible" aporias—themes in which the condition of their possibility is also, and at once, the condition of their impossibility. Aporias are hence at the heart of Derrida's quasi-transcendental philosophy, being both enabling and disabling conditions. But we should also note, with a view to the previous chapters that have examined Derrida's transcendental philosophy, that aporias are argued to be both conditions for anything to appear, as well as regulative ideals for anything that merits the name of genuine thinking and creation: for example, "go there where you cannot go, to the impossible, it is indeed the only way of coming or going."[4] These kind of quasi-transcendental comments again have a dual register, admitting both metaphysically inclined interpretations as well as more modest conceptual interpretations, and Derrida adds that the point is not so much to get out from or escape an aporetic impasse, but to invent "another thinking of the aporia, one perhaps more enduring" (A 13).

The first aporia that Derrida treats is in fact Aristotle's temporal aporia. We want to say both that the "now" is, and that it is not what it is, but to cut a long story short, for Aristotle (as with Zeno, McTaggart, etc.) it is impossible to maintain time as both an entity and not an entity simultaneously, due to the contradictions that ensue. Of course, the aporia that Derrida is concerned with is better captured by the Epicurean slogan about death: where death is, I am not; where I am, death is not. To paraphrase: as long as a subject exists, it has not yet reached its end; when the end has come, there is no subject around to be harmed by this. The "not yet," of course, shows the centrality of temporal experience to any human subject, who exists essentially as engaged in projects that have a future orientation and is not yet at their end. In this respect, the aporia of death is also famously expressed by Heidegger in *Being and Time*. Heidegger states: "death, as the end of *Dasein*, is *Dasein*'s ownmost possibility" (BT §52). He then adds: "the more unveiledly this possibility gets understood, the more purely does the understanding penetrate into it as the possibility of the impossibility of any existence at all" (BT §52). Death is, then, the possibility of an impossibility. In fact, Heidegger contends that *Dasein*'s most proper possibility is this possibility of an impossibility. Is this a logical contradiction, paradox, aporia? Can we say this while respecting logic and meaning? Can we think this statement, which seemingly elides common sense? Do we understand it as the possibility of impossibility as Heidegger maintains, or as the impossibility of possibility as Levinas contends in *Time and the Other*, a Byzantine distinction that Levinas maintains (and Derrida agrees) is actually highly significant? But let us run with this enigmatic claim of Heidegger's, at least for the time being, so as to bring to light tensions within the text of *Being and Time*, and places at which Derrida contends that the arguments of this great book begin to undermine themselves.

Ontology of Death Versus Ontic Treatments of Death

It is worth noting that Heidegger's overarching concern in *Being and Time* is to revisit the question of the meaning of Being, a project that depends upon an equally famous if sometimes opaque distinction, that he calls the ontic-ontological difference. Roughly, this can be schematized as follows:

Ontical enquiry examines entities, or beings (particular treatments of death)

Ontological enquiry examines that which allows entities to be, or Being (death per se, the condition of possibility of particular treatments)

Heidegger's general claim is that all ontic inquiry, including science, is said to presuppose an ontology. Fundamental ontology grounds the various regional ontologies, which in turn ground the specific sciences. As would be apparent, these are orders of presupposition and priority, transcendental claims. Moreover, on this Kantian-inspired view philosophy remains the queen of the sciences.

In his famous account of being-towards-death, Heidegger also wants to foreground a difference of this order between his existential analytic of death and what non-philosophical approaches to death involve. He does this by radically distinguishing dying from perishing and demising. These careful distinctions and orderings depend upon, and circularly help to shore up, the pivotal ontic-ontological distinction itself. Bound up with these terms is also a distinction between an authentic relationship to death, in which one resolutely recognizes that death is our ownmost possibility, and an inauthentic relationship to death, in which one fears death as an event (i.e., the prospect of one's own demise), or one is indifferent to death (since one might treat their end as something that comes naturally to us all).

For Heidegger, all ontic enquiries into death, which study death in relation to a particular domain are said to presuppose the meaning of death per se. Heidegger hence puts in play a logic of presupposition. All the disciplines, with their regional borders (e.g., demography), necessarily presuppose a meaning of death, a preunderstanding of what death is, and in his existential analysis of death Heidegger seeks to make this preunderstanding explicit and to explain it (that is to ground it, to offer their condition of possibility). Theology, anthropology, history, biology, and so on are all said to presuppose something that does not let itself be addressed within these specific domains, and naively put into play conceptual presuppositions about life and death. Biological accounts of death might, for example, treat it naturalistically in regard to biogenetics and the cessation of lifespan. But they also presuppose a concept of death, the meaning of death in general, in a manner that is not restricted to their specific ways of treating it (A 44). Of course, any such naivety is entirely appropriate. Ontic accounts of death need to have a presupposed understanding of death so that they know what it is they are examining and can limit their enquiries in appropriate ways, and so that they can recognise when they have found what they are looking for (as with Meno's learning paradox). But questions regarding the meaning of death in general, what death is, are sidelined, and for Heidegger it is only fundamental ontology, transcendental philosophy, that examines death as such without sacrificing the question too quickly in the search for answers and for applications. In other words, such disciplines all presuppose philosophy, or for Heidegger, they presuppose onto-phenomenology, which "does not let itself be enclosed within cultural, linguistic, national, or religious borders either, and not even within sexual borders" (A 27). As Derrida says, summarizing Heidegger:

> The historian knows, thinks he knows, or grants to himself the unquestioned

knowledge of what death is, of what being-dead means: consequently, he grants to himself all the criteriology that will allow him to identify, recognise, select, or delimit the objects of his inquiry or the thematic field of his anthropologico-historical knowledge. The question of the meaning of death and of the word "death," the question "What is death in general?" or "What is the experience of death?" and the question of knowing if death *is* . . . all remain radically absent *as questions* (A 25).

In relation to theological accounts of death, Heidegger notes that the question of the after-life only has meaning if one has first elaborated a concept of the ontological essence of death (A 53). In relation to psychological and anthropological accounts of death, such analyses treat death in terms of culture and history. They might seek to describe cultural difference in attitudes toward death, noting, for example, that some cultures eat the dead, whereas some put them in graves or cremate the deceased. But all such ontic accounts of rituals concerning death are, Heidegger suggests, ceremonial rather than ontological, and he insists that our certainty of death is heterogenous to any other certainty, whether theoretical or empirical, including the spectacle of the other's demise. This certainty, which the anthropologist must take for granted in their enquiries, cannot be explained on such grounds. How does Heidegger justify this kind of argument? It is a crucial series of distinctions that allow Heidegger to claim that the existential analytic of death is before any metaphysics of death, anthropology of death, psychology of death, and biological account of death (A 27). He will need to preserve a domain, a transcendental domain, in which the existential analysis is not only first, but it is also neutral in regard to all of the above (and hence nature and culture). As such, philosophy (properly done in a Heideggerian manner) is the only discipline for which an analysis of death has no border, no limit, and yet the analysis of which creates a distinct border between that which is *Dasein* and all that is not *Dasein*, including the animal, which in *The Fundamental Concepts of Metaphysics* he famously called poor in world, *weltarm.*[5]

Perishing, Demising, and Properly Dying

Heidegger draws a distinction between three ways of relating to death. Ontic accounts of death are concerned with perishing or demising, but only an ontological account of death is concerned with dying as such, properly dying. Perishing, then, refers to the end of the living, and the prospect of "kicking the bucket" is something that all living things share. Perishing is thus largely a biological term. *Dasein*, Heidegger repeatedly says, never perishes, although the human biological form obviously does (the term *Dasein* refers to being-there, our characteristic mode of being-in-the-world). This is not properly dying for Heidegger, and he makes a related point in regard to demise. *Dasein* alone can demise in the medico-legal sense, after biological persishing. Demise is thus characteristic of *Dasein*, but it is still not properly dying. As Derrida observes: "Demising is not dying but, as we have seen, only a being-toward-death . . . can also demise. If it never perishes as such, as *Dasein* (it can perish as a living thing, animal, or man as rational animal, but not as *Dasein*) . . . *Dasein* can nevertheless end, but therefore end without perishing and without properly dying" (A 39). Demise, we might suggest, is an attitude toward this biological fact of perishing that Heidegger alleges only humans can have (Derrida disagrees), but it is an attitude that is insufficiently ontological and thus inauthentic, since it denies the structure of *Dasein*. This is because in the kind of relationship to death characteristic of demise, we treat death as an external event (an ontic event) that will eventually happen to us and cause us to cease to be (and typically we resent or fear this). In demise we are viewing death from the outside, so to speak, even if it pertains to us, rather than seeing being-towards-death as fundamental to who we are right now (properly dying).

On this latter view, (properly) dying is an existential structure that defines *Dasein*, and this means that the possibility of dying is part of the structure of our world as we experience it now, not just something that is deferred until later. We might say that death is a futural possibility that is constitutive of the now of the present; to phrase it in the first-person as Heidegger intimates we must, my present is what it is, only owing to my understanding that this present is finite and that it will not go on forever. In other words, my understanding of a final "not yet," a final possibility, is what allows me to meaningfully structure and organize my life. Awareness that I am going to die allows me to get a perspective on life as a whole, and to establish the resoluteness that Heidegger associates with authenticity, because, as David Farrell Krell puts it, it "invades my present, truncates my future, and monumentalises my past."[6]

While Heidegger claims that properly dying is *Dasein*'s ownmost possibility that no one else can undergo for us (BT §52), Levinas in various texts asks us to consider what is our first and enduring experience of death. He suggests that it is the other's death, or the prospect of the other's death, rather than the inevitability of our own. But, for Heidegger, this is an ontic claim rather than an ontological one. As such, when Levinas says that the death of the other is the first death metaphysically speaking, the Heideggerian rejoinder is that Levinas hereby invokes the experiences that we have of the death of the other in their *demise*, which are not of the same significance as one's own experience of dying, either existentially or ontologically (A 39). In regard to this dispute, we saw in chapter 8 that Derrida thinks that the alternatives here are somewhat superficial, but like Levinas he is concerned that Heidegger's account seems to render mourning of the other's death (as well mourning of the other that is part of the self and yet different) as philosophically inconsequential. He suggests that, "the existential analysis does not want to know anything about the ghost or about mourning" (A 60), and we have seen that both tropes are, by contrast, fundamental to Derrida's later work. In *Spectres of Marx*, for example, the ghost is neither present nor absent, or it is both present and absent, and the further suggestion is that such undecidable structures are fundamental to what has meaning for us. And to return this discussion to our overarching concern with time and politics, we might note that the past is ghost-like, the future is ghost-like, not present but not quite absent either. Refusing to mourn these dimensions of temporal experience, deciding in favor of the resolute anticipation of death and against the ambiguity of temporal experience, is, however, a decision with ethico-political significance for Derrida, and although he does not dwell on this at any length in *Aporias*, it is difficult not to think that Derrida has Heidegger's infamous allegiance to National Socialism in mind.

Indeed, like Levinas, Derrida is concerned with Heidegger's emphasis on "properly dying." The authentic being-able of *Dasein* might be said to a discourse of mastery and virility over death: of being able, capable, to resolutely anticipate death. That is why it is significant that death, for Heidegger, is claimed to be the "possibility of an impossibility" rather than the "impossibility of possibility." But Levinas responds to Heidegger (without naming him in this passage):

> The ancient adage designed to dissipate the fear of death—"If you are, it is not; if it is, you are not"—without doubt effaces our relationship with death, which is a unique relationship with the future, But at least the adage insists on the eternal futurity of death . . . The *now* is the fact that I am master, master of the possible, master of grasping the possible. Death is never now. Where death is here, I am no longer here, not just because I am a nothingness, but because I am unable to grasp. My mastery, my virility, my heroism as a subject can be neither virility nor heroism in relation to death."[1]

In different ways, then, both Levinas and Derrida accuse Heidegger of this heroisim in relation to death, but it is Derrida who best reveals the peculiar philosophical tangles that Heidegger's discourse of the "good death" is stuck in, Heidegger's own barred path.

The Animal That Therefore I Am

Of course, the issue of animality is central to *Aporias*, as it is to much of Derrida's work. To see its signficance for this dispute let us recall that, for

Heidegger, disciplines other than philosophy all also allegedly treat death ontically, more particularly as demise. As Derrida puts it:

> On the one hand, there would be anthropological problematics. They would take into account ethnologico-cultural differences affecting demise, sickness, and death; however, on the other hand, and first of all, there would be the ontologico-existential problematic that anthropology must presuppose and that concerns the being-until-death of *Dasein*, beyond any border, and indeed beyond any cultural, religious, linguistic, ethnological, historical and sexual determination. In other words . . . there is no culture of death itself or of properly dying. Dying is neither entirely natural (biological) nor cultural. (A 42)

Certainly, for Heidegger anticipating death is the preserve of *Dasein*, and, although they are not reducible, only the human is *Dasein*. The animal, the living thing, is not mortal in the sense that *Dasein* is for Heidegger; they cannot comprehend finitude and anticipate death. An animal can merely come to an end, perish. Derrida suggests that this is empirically wrong, claiming that animals have a sense of demise, an interpretation or cultural response to biological death of a sort, and he also points out that neither *Dasein* nor animals have a direct relation to death through dying (proper). Even language may dissimulate rather than reveal death. Derrida comments:

> Animals have a significant relation to death, to demise and thus to mourning. They might not have a direct relation, but nor do we . . . Maybe both humans and animals never have a relation to death as such, but only to perishing, demising, the death of the other, even if it is the other within, death of the other in me. (A 76)

As would be apparent, there is a strong distinction between *Dasein* and all else in *Being and Time* (and in later texts), even where Heidegger purports to have left behind defining the human in terms of essential traits (e.g., rational animal). It also seems to follow that if the living thing as such (e.g., the animal), does not have anything to do with death as such (proper death), then nature and life have nothing to do with death, and we have a series of oppositions that Derrida will seek to problematize: life versus death; culture versus nature, etc. Indeed, Derrida begins the second essay of this book by stating that, "the difference between nature and culture, indeed between biological life and culture, and, more precisely, between the animal and the human is the relation to death . . . Although Heidegger, deeply rooted in this tradition, repeats it, he also suggests a remarkable rearticulation of it" (A 43-4). Of course, Heidegger would maintain that there is neither a nature nor a culture of being-towards-death (A 42), and it follows that if an analysis of being-towards-death is neutral on culture it will also be neutral on politics. As such, there can be no politics of death properly speaking for Heidegger (A 59). Derrida and Levinas, however, both dispute that this is so. On their view, Heidegger surreptitiously has a politics of death, of what properly dying consists in (A 60), despite the fact that his neutral transcendental order of priority should have no truck with it. And on Derrida's view this is not a superficial accident, a mistake on Heidegger's behalf that might have been avoided had he been more careful. Rather, as Derrida puts it, "there is no politics without an organization of the time and space of mourning" (A 61), and Heidegger's "failure" is not to recognize this. To separate politics from pure ontology in the manner that Heidegger claims to is untenable. Heidegger's discourse is not politically neutral, and it does give us a moral discourse on the good death involving tropes like mastery and virility as Levinas suggests. Indeed, Derrida concludes, with Levinas, that, "a certain thinking of the possible is at the heart of the existential analysis of death" (A 62). Moreover, it is important to recognize that this possibility is not simply described, but it must also be assumed in the resolute anticipation of being-towards-death. Derrida hence observes that "the statements of the existential analysis are originally prescriptive or normative" (A 64), contrary to what Heidegger would hold: his allegedly neutral transcendental philosophy of time and death is always more than it claims to be, always involving moral and empirical claims that it had attempted to dispel.

A Principle of Contamination

Derrida's claim, then, is that these three carefully defined "endings"—perishing, demising, and properly dying—bring together all of the paradoxes and chiasmi that both relate and unravel Heidegger's existential analysis of death. As he comments:

> The distinction between perishing and dying has been established, as far as Heidegger is concerned, as he will never call it into question again . . . . if, in its very principle, the rigour of this distinction were compromised, weakened, or parasited on both sides of what it is supposed to dissociate, then the entire project of the analysis of *Dasein*, in its essential conceptuality, would be, if not discredited, granted another status that the one usually attributed to it (A 31-32).

Now I have already noted the indebtedness of my own book to Heidegger's analyses of the ready-to-hand and pragmatic temporality, both of which come from the existential analytic of *Being and Time*. I hence do not myself agree with Derrida that the entirety of Heidegger's book is jeopardized on the basis of his acute deconstruction, although several of its governing distinctions are: the authentic-inauthentic distinction, for one, as well as certain readings of the ontic-ontological difference. Derrida highlights the overreaching of *Being and Time*, a chronopathological tendency in which it exceeds its own methodological strictures and stems from elsewhere than ontological necessity or phenomenological demonstration. Heidegger's distinction between dying, demising and perishing, "are threatened in the very principle, and, in truth, they remain impracticable as soon as one admits that an ultimate possibility is nothing other than the possibility of an impossibility" (A 77). But also, they are contaminated by biology, by the other's demise and its significance for us (whether imagined, projected, or actual), and by anthropology, theology, and culture more generally, which mean that any ontology is never pure and neutral. Derrida is rigorously attentive to this tendency of philosophy, this anteriority complex as Alan Murray calls it,[8] in which philosophy attempts to purify itself, to "oppose rigorously two concepts or the concepts of two essences, and to purify such a demarcating opposition of all contamination, of all participatory sharing, of all parasitism, and of all infection" (A 41). An attempt that breaks down, as Derrida shows, as the various distinctions are problematized, not only by the very paradoxicality and ambiguity enshrined in stating that death is "the possibility of an impossibility," but also in the various Judeo-Christian cultural assumptions in Heidegger's purportedly neutral (transcendental) account of death, notably the great distance between man and animal, but also the themes of guilt, conscience, fallenness, etc., all have a moral element, as does the authentic-inauthentic distinction and the ontic-ontological distinction itself. Precisely where Heidegger claims he has overcome this, we can see he is still within a history of cultural accounts of death, not neutrally universal and a priori, and yet his various distinctions do a lot of work guarding philosophy's proper domain, and even serving to guard against other potential philosophical replies, like Levinas' and indeed like Austin's reply to Marcel with which we began this chapter. As such, they are both enabling and disabling conditions for *Being and Time* as a whole.

Of course, Derrida does not mean to simply critique Heidegger—deconstruction is rarely as simple as critique—but to show the manner in which the discipline of philosophy itself is revealed as aporetic, not reducible to other disciplines, but also not able to have its own proper domain that is divorced from other disciplines. Instead, we have what Derrida calls a principle of contamination. This does not mean the end or death of philosophy as was occasionally intimated by some of the poststructuralists in the late 60s, but it does problematize some of the metaphilosophical distinctions that philosophers have typically used to define their terrain (e.g., ontological vs. ontical inquiry in Heidegger), and it does problematize any anteriority complex that philosophers might have, be they continental or otherwise. Instead of ontology, what we have, as Derrida says in *Spectres of Marx*, is a hauntology, in which philosophy is always haunted by other disciplines, never able to secure its own domain once and for all. For Derrida, though, the fact that transcendental philosophy is also intrinsically aporetic does not mean we can do philosophy without transcendental arguments. Although the projecting of grounding and rigorously distinguishing between the empirical and the transcendental inevitably breaks down, this itself is a quasi-transcendental law. Moreover, it has been argued throughout parts 2 and 3 of this book that there is a related "law" of mutual contamination between the phenomenological and poststructuralist accounts of time. In the next and final chapter, the conclusion, I attempt to make this as clear as possible and to extend the conversation back to analytic philosophy and consider again some of the metaphilosophical strengths and weaknesses of these approaches to time and temporality.

13

Conclusion: Beyond Chronopathologies

If it is not already clear, I think that if anything warrants the name of *aporos* it is the state of contemporary philosophy, and the methodological and topical divergences surrounding time that I have attributed to, respectively, poststructuralism, phenomenology, and analytic philosophy. The tension between these competing philosophical norms is especially apparent in regard to analytic and poststructuralist philosophy, but the divide between phenomenologists and poststructuralists is also quite entrenched and recalcitrant, albeit without mutual ignorance being the norm. Indeed, to see that this is so one needs only to consider the polarizing views about whether or not Derrida's early engagements with Husserl in *Speech and Phenomena* and *Introduction to Husserl's Origin of Geometry* constitute forceful objections to the phenomenological project or are willful misreadings that do not touch the Husserlian project in any meaningful way.[1] But is this really an aporia? It seems to me that it is, since these three major theoretical trajectories all have metaphilosophical, methodological, and normative justifications bound up with them that are often incompatible. Internalizing the norms of any of these traditions makes taking the other(s) seriously very difficult, and it is hence exceedingly difficult to say, in an optimistic and pluralist fashion, let a thousand flowers bloom.

Allow me to draw together some of the main threads from this book that justify such a claim. In part 1 of the book we saw evidence of a chronopathology prevailing in analytic philosophy: an atemporality in some cases (e.g., utilitarianism, four-dimensionalism, etc.), a presentism in other cases (we have to start from somewhere: e.g., the intuitions of the "folk"), both tendencies that were apparent in John Rawls' liberal philosophy. In addition, in political philosophy there is also a general preoccupation with distributive justice that remains anathema to most forms of continental ethico-political reflection. This chronopathology is also ensconced in some of the core philosophical methods of analytic philosophy more generally: analysis aims to reveal underlying structures (sometimes logical) that are timeless; truth is defined as independent of all justification (and hence historical processes); thought experiments are "intuition pumps" that call on pretheoretic opinion and then structure debates by serving as placeholders; the deferential relationship to the best findings of the sciences indexed to the present (hence an expert presentism, which requires perennial updating); the *prima facie* credence given to common sense (a nonexpert presentism); as well as the general prevalence of coherentist devices like reflective equilibrium which take seriously both pretheoretic opinion and the findings of the various expert domains.

To suggest that this amounts to a time-sickness might seem to beg the question by assuming that such views are false, and it might be fairly protested that nothing quite so strong has been established in this book. But my weaker claim here is simply that the methods and metaphilosophical understanding of the role of the philosopher are temporally limited and partial. I would also add, however, that any claims to them being exhaustive seem especially weak in regard to ethico-political reflection. Of course, any reservations about analytic philosophy raised by continental philosophers that is historicist in motivation, or that refers to a conjunction of time and politics, will not typically engage analytic philosophers and cause them to worry much about their philosophical presuppositions. One fairly typical analytic response is Hans-Johann Glock's in *What Is Analytic Philosophy?* He suggests that any claim about the historical presuppositions of concepts and theoretical frameworks is either trivially true if intended in a weak sense (and not incompatible with analytic practice), or if intended in a strongly historicist sense it is false (i.e., a pernicious relativism).[2] For most continental philosophers, however, it is not simply trivially true, but deeply true, in the sense that this is what all philosophy ought to involve. Whether this commits them to relativism is not something that can be explored here, but we might recall that, for Hegel, philosophy is essentially about grasping one's own time in thought, and for Deleuze and Guattari (and Foucault) philosophy is about furnishing theoretical resources to help us to think otherwise and hence resist the hegemony of the present. In this respect, of course, we noted in the introduction that Putnam objected that continental philosophers give us a "parapolitics," a militarized philosophy, and this kind of concern has been at the heart of the analytic reception of continental philosophy since the distinction began to have genuine socio-political force in the context of the two world wars.[3] Without directly addressing the substantive aspects of Putnam's charge here, the attribution of parapolitics is, I think, mistaken, since in my view the main thrust of thinkers like Deleuze and Derrida is actually an attempt to imbue the political with the ethical. They aim to interrupt and problematize any too easy calculations in the name of the singularity of the other, the difference of the future, and to remind us of the way that methodologies that focus upon such calculations level off the past and the future, as well as the affective dimensions of problems more generally.

Of course, no salient political philosophy (and arguably philosophy more generally) can afford to do without either of these trajectories, the analytic and poststructuralist. In sensible policy making, for example, we will have to consider present aims and preferences, and what would be a just distribution of goods. But Derrida and Deleuze are right to suggest both that the ethico-political (as opposed to politics, more narrowly construed) cannot be reduced to this, and that the motivation and impetus for change generally requires both a more utopian (future-oriented) and a more critical and genealogical (past-oriented) dimension. Moreover, both of them suggest that one important way of thinking about the difference between ethics and politics is to recognize that it constitutes a temporal difference, a rhythmic difference (R 60). This is, for example, why Deleuze insists on the competing and non-synthesizable demands of the times of Chronos and Aion, differing rhythms and times that we might venture to suggest are the times of politics and ethics respectively (the orders of political calculation and ethical incalculability, the latter including a certain utopian dimension). Despite these important insights with which I am largely in agreement, I have attempted to show in this book, especially in part 2, that Deleuze and Derrida accord far more attention to ethical time (which is incalculable and split open in the direction of the immemorial past and the future that cannot be anticipated) than political time (which is calculable and presentist).

In my view, however, we cannot rest content with this trajectory since its great strength is also its weakness. The attempt to articulate conditions for present experience and present identity tends, apparently inexorably in the work of the poststructuralists, toward a prophetic futurity and a philosophy of difference that involves, more or less directly, a denunciation of good sense and calculative time. If I am right about the general contours of this position, this suggests that a rapprochement between analytic political philosophy and poststructuralism (and continental philosophy more generally) is desirable. Whether it is genuinely possible (for philosophical as much as socio-political reasons) remains less clear given that some fundamental metaphilosophical and normative orders of priority make it difficult for the twain to meet. This is due to norms like methodological empiricism and respect for common sense in analytic philosophy, and the continental "temporal turn" with its association with a hermeneutics of suspicion and transcendental philosophy. Hence the aporia: while each tradition needs the other, each is also precluded from that very dialogue to the extent that they take these norms seriously, especially in regard to the divergent methodological and topical significance given to time. The aporetic challenge for our time(s) is hence to do what we as philosophers find almost impossible to do: engage in genuine conversations with our respective others, whether they be analytic, continental, phenom-enological, or poststructuralist. To put it more Socratically, we need to know that we do not know, which as with the Greek use of aporia, ought to inspire us to find out more. Such a desire is a precondition for any postanalytic and metacontinental future to emerge.[4]

Hopefully this book has made a contribution to creating such a desire, or at least to providing further resources to grapple with, for those already gripped by such a desire. If it has, then we might be entitled to be just slightly more optimistic than pessimistic about the aporetic situation of contemporary philosophy. Indeed, any aporia also has its possible solutions, even if the solutions will inevitably be partial, and even if the aporetic problem is more important than the solution as Deleuze and Derrida both suggest. Derrida, for example, says: "when someone suggests to you a solution for escaping an impasse, you can almost be sure that he is ceasing to understand, assuming that he had understood anything up to that point" (A 32). Deleuze maintains that "a solution always has the truth it deserves according to the problem to which it is a response" (DR 159), and he adds that "problems are the differ-ential elements in thought, the genetic elements in the true" (DR 162). These comments are acute, but they do not discharge the philosophical obligation of providing solutions as best one can and of intervening. Moreover, as would be clear from the foregoing, I do not accept the Deleuzian understanding of philosophy as fundamentally about concept creation rather than the model of judgment, nor the too strident disassociation of philosophy from common sense. As such, here are my tentative solutions/interventions, which certainly do not exhaust the problematic in question.

First, I think the history of philosophy shows that claims to transcenden-tal priority concerning time are typically accompanied by illegitimate (in the sense that they cannot establish their necessity) normative moves, whether in Husserl, Heidegger, Derrida, Deleuze, and other philosophers not considered in any detail here. One response to this might be to consign transcendental philosophy to the dustbin of history. But from my perspective the best solu-tion is not to give such reasoning up, since the consequence seems to be an impoverished account of time and politics that is itself partial and dogmatic in various different ways. In particular, it cannot offer an "ontology of the present" as Foucault puts it, a genealogy of ourselves, and instead seems committed to an intellectual view that Simone de Beauvoir describes as the curator of the given world.[5] If the recent history of contemporary thought at-tests to the manner in which various temporal assumptions are bound up with metaphilosophical priorities and orders of presupposition, there is reason to think that we need to be more attentive to the manner in which each of these serves as a limit to the other(s). Perhaps the best explanation is one in which transcendental reasoning and the project of grounding is viewed as necessarily limited by each of the transcendental philosophies of time associated with phenomenology and poststructuralism, but one that cannot be pragmatically dispensed with either. The quasi-transcendental law that we might infer, then, is one regarding the irreducibility of a movement toward equilibrium associated with the living-present to invoke Husserl and Merleau-Ponty (or associated with the experience of time as a unity, to invoke Heidegger and authentic *Dasein*), as well as that which breaks this open and precludes it being attained once and for all. And it is not that the one is somehow good, and the other bad (or reactive, slavish, illusory, etc.). Both have an ethico-political significance, and the aporetic and dialectical demand of philosophy is to remain attentive to each.

In this respect, my solution is decidedly Aristotelean. Consider Aristotle's doctrine of the mean and virtues in the *Nichomachean Ethics*: "So virtue is a purposive disposition, lying in a mean that is relative to us and determined by a rational principle, and by that which a prudent man would use to determine it. It is a mean between two kinds of vice, one of excess the other of deficiency."[6] What I am suggesting here, in effect, is that intellectual and theoretical virtues obey a similar logic. Philosophy needs to be able to adequately come to terms with what Hoy calls the *time* of our lives—I prefer to pluralize this to the *times* of our lives—and yet the poststructuralism of Deleuze and Derrida has an excessive emphasis on times (and methods) that resist the present while simultaneously at work within the present, and analytic philosophy tends to have a deficiency of such methods generally being either atemporal or presentist. What is needed, it seems to me, is the "golden mean" as Aristotle calls it, and which I suggest that a phenomenology of the lived body can help to provide. In their structural coupling with a given environment, our proprioceptive bodies inevitably seek to find a mean (equilibrium) in any situation, and this kind of pragmatic temporality that is bound up with the constitution of the living-present also has an ethical significance, with Dreyfus associating it with an ethics of phronesis and practical wisdom. This may sound old-fashioned and unexciting when counterposed with a metaphysics of the different and the new, but such a view makes possible something tantamount to a middle way from which we can navigate between the different symptoms endemic to analytic and continental philosophy, guarding against the various excesses and deficiencies that have been delimited in this book. In particular, it enables one to avoid any privileging of calculable rationality, which characterizes large swathes of analytic philosophy (including the unerring focus on distributive justice), without lapsing into a prophetic politics of the incalculable in which judgment and calculation, and the structures of temporal experience that sustain them, is maligned. Any calculative presentism, for example, involves a restricted aspect of the pragmatic time of the living-present but it is not a priori beside the point for phenomenology. Phenomenology hence need not be as metaphysically and methodologically committed as analytic philosophy as a community tends to be, nor as the transcendental philosophies of the future of Deleuze and Derrida are, at least in their worst moments when the interconnection between time and transcendental philosophy is used to denounce certain doxic features of common sense and the centrality of any more pragmatic and embodied conception of time. Both views too readily lay claim to having attained to *the* philosophical stance. As such, it should also be clear that my call for a return to the pragmatic temporality of phenomenology (including the living-present) should not be understood to be be a call to phenomenology as first philosophy in any foundationalist fashion, but as involving a twofold strategy: an argument for the irreducibility of this embodied perspective and the proffering of some pragmatic reasons for starting here rather than elsewhere. In particular, such a view allows for a synthesis of many of the best aspects of analytic philosophy and poststructuralism. Although some might not share this view, I think that this is a reasonably significant factor, albeit a contingent one, in favor of the view I advance. Such a view presupposes that there is something (philosophically) valuable and useful in all of the three trajectories delineated in this book, but it maintains that the best way to keep all of the methodological balls juggling in the air so to speak, and to recognize the virtues of each without succumbing to their problems, is to start with the phenomenological time(s) of our lives.

Notes

1. Introduction: The Politics of Time

1. Jeremy Rifkin, *Time Wars: The Primary Conflict in Human History* (New York: Touchstone, 1987), 10.

2. Ludwig Wittgenstein, *Philosophical Investigations*, ed. G. E. Anscombe (London: Blackwell, 1953). See §65-6, especially the famous discussion regarding the definition for a game—no necessary and sufficient condition definition seems to suffice—and the rope that has a unity despite no single thread persisting throughout the whole rope.

3. Elizabeth Grosz, "Thinking the New: Of Futures Yet Unthought," in *Becomings: Explorations in Time, Memory and Futures*, ed. Elizabeth Grosz (Ithaca, NY: Cornell University Press, 1999), 18.

4. David Hoy, *The Time of Our Lives: A Critical History of Temporality* (Cambridge, MA: MIT Press, 2008).

5. Quentin Meillassoux, *After Finitude*, trans. Ray Brassier (London: Continuum, 2008). Meillassoux contends that many continental philosophers are "correlationists" to either a weak or a strong degree (depending, roughly, on the extent to which their inheritance is Kantian and Hegelian), and so cannot adequately account for the reality of ancestral claims regarding "arche-fossils" (including in physics, biology, etc.) that predated human subjects.

6. Hoy, *The Time of Our Lives*, 22.

7. In *Analytic Versus Continental: Arguments on the Methods and Value of Philosophy* (Durham, UK: Acumen, 2010) James Chase and I argue that some of the key indicia of the continental tradition include such matters as: a wariness about aligning philosophical method with commonsense, a "temporal turn" which encompasses both ontological issues and an emphasis on the historical presuppositions of concepts and theoretical frameworks, an interest in thematizing inter-subjectivity, an anti-representationalism about the mind (including what we today call computational and functionalist models), an investment in transcendental arguments and more generally transcendental reasoning, a concern with the relationship between style and content, a critical and nondeferential (or transformative) attitude to science, resistance to mechanistic or homuncular explanations (say, in regard to science and philosophy of mind: this is tied to anti-representationalism), and an antitheoretical attitude to ethico-political matters (no formalizing of ethics for decision procedures).

8. David Cooper, "Nietzsche and the Analytic Ambition," *Journal of Nietzsche Studies* 26, No. 1 (2003): 2.

9. In this respect, see Simon Glendinning, "Argument All the Way Down: The

Demanding Discipline of Non-Argumento-Centric Modes of Philosophy," in *Postanalytic and Metacontinental: Crossing Philosophical Divides*, eds. Jack Reynolds, James Chase, James Williams, and Edwin Mares (London: Continuum, 2010), 71-84.

10. This divide seems to partly stem from when it becomes clear that coincidence that Davidson is doing transcendental philosophy. The most cited papers in *Mind* and the *Journal of Philosophy* are the ones before Davidson's project clearly becomes a transcendental one: "Actions, Reasons and Causes," "Truth and Meaning," and "Mental Events." In more crossover journals, like *European Journal of Philosophy* and *Southern Journal of Philosophy*, it is essays after "On the Very Idea of a Conceptual Scheme" that are more frequently cited. For more on this in regard to so-called postanalytic philosophers like Davidson, McDowell, Rorty and Wittgenstein, see George Duke, Elena Walsh, James Chase, and Jack Reynolds, "Postanalytic Philosophy: Overcoming the Divide?" in *Postanalytic and Metacontinental: Crossing Philosophical Divides*, eds. Jack Reynolds, James Chase, James Williams, and Edwin Mares (London: Continuum, 2010), 7-24.

11. See Barry Stroud, "Transcendental Arguments," *Journal of Philosophy* 65 (1968): 241-56; Stephan Körner, "Transcendental Tendencies in Recent Philosophy," *The Journal of Philosophy* 63 (1966): 551-61; and Stephan Körner, "The Impossibility of Transcendental Deductions," *The Monist* 51, No. 3, July (1967): 317-31.

12. This is James Chase's idea, and we develop it in *Analytic Versus Continental*.

13. Steven Crowell, "The Project of Ultimate Grounding and the Appeal to Intersubjectivity in Recent Transcendental Philosophy," *International Journal of Philosophical Studies* 7, No. 1 (1999): 31.

14. While Foucault comments in an interview that he attempts to "historicise to the utmost in order to leave as little space as possible to the transcendental," this statement indicates that it is an effort that in his view can never totally succeed. See Michel Foucault, *Foucault Live*, ed. S. Lotringer, Trans. L. Hochroth and J. Johnston (New York: Semiotext(e), 1996), 99.

15. Michel Foucault, "The Art of Telling the Truth," in *Critique and Power*, ed. M. Kelly, trans. Alan Sheridan (Cambridge, MA: MIT Press, 1994), 147-48.

16. Mark Sacks, "The Nature of Transcendental Arguments," *International Journal of Philosophical Studies* 13, No. 4 (2005): 444.

17. Simon Glendinning, "Reply to Reynolds," *International Journal of Philosophical Studies* 17, No. 2 (2009): 273-80. Also, see his book, *The Idea of Continental Philosophy* (Edinburgh: Edinburgh University Press, 2006).

18. Brad Hooker, Elinor Mason, and Dale Miller, eds. *Morality, Rules and Consequences: A Critical Reader* (Lanham, MD: Rowman and Littlefield, 2000), 1.

19. Friedrich Nietzsche, *Beyond Good and Evil*, trans. Walter Kaufmann (New York: Random House, 1966), §253.

20. Timothy Mulgan, *Understanding Utilitarianism* (Chesham, UK: Acumen, 2007), §2.

21. See Iris Marion Young, *Justice and the Politics of Difference* (Princeton: Princeton University Press, 1990); Bonnie Honig, *Political Theory and the Displacement of Politics* (New York: Cornell University Press, 1993); and Michael Walzer, *Spheres of Justice* (New York: Basic Books, 1984).

22. Young, *Justice and the Politics of Difference*, 8, 11, 16.

23.   Hilary Putnam, *Renewing Philosophy* (Cambridge, MA: Harvard University Press, 1992), 197.

24.   David Wood, *The Step Back: Ethics and Politics after Deconstruction* (Albany, NY: SUNY Press, 2005), 32.

2. Analytic and Continental Philosophy: A *Contretemps*?

1.   This chapter was originally written as a public talk to the Melbourne Writer's Festival in 2008, and a shorter version is being published by Graham Oppy and Nick Trakakis, ed. *The Antipodean Philosopher, vol. 2: Public Lectures on Philosophy in Australia and New Zealand* (Lanham, MD: Lexington Books, 2010). An expanded version, co-written with James Chase, is also a chapter in our book, *Analytic Versus Continental*.

2.   See Bernard Williams, *Truth and Truthfulness: An Essay in Genealogy* (Princeton, NJ: Princeton University Press, 2002), 4; and Hilary Putnam, "A Half Century of Philosophy," *Daedalus* 12 (1997): 203.

3.   It also challenges the "growing universe" theory, which holds that both the present and the past are real (the name deriving from the idea that any list of entities that are real necessarily increases over time as things that are present become past), but we cannot consider this position here.

4.   John M. McTaggart, "The Unreality of Time," in *The Philosophy of Time*, Eds. Robin Le Poidevin and Murray MacBeath (Oxford: Oxford University Press, 1993), 23-34.

5.   Russell published a strongly worded critical review of Bergson's work, "Philosophy of Bergson" (*Monist* 22, 1912: 321-47), and continued to criticize his colleague from the other side of the channel for years to come, and the disparity between their views on time and multiplicity became apparent when they were co-participants at a conference with Einstein in 1922 at the Collège de France. Bergson also gets a tough time in Russell's *History of Western Philosophy* (London: Routledge, 2004).

6.   See Russell, "Philosophy of Bergson." Bergsonian intuition is understood by Russell to be equivalent with mere instinct, which is not quite true; it develops from instinct, but this does not mean it is reducible to it. And certainly Bergson does not think the instinctive abilities of bats and bees allow them to do metaphysics! Nonetheless, it seems fair to say that there is an anti-intellectualist spirit in the work of Bergson that can be discerned in much of twentieth century continental thought that the analytic and Russellian projects do not share. And it is, of course, paradoxical for philosophers to take positions that seem anti-intellectualist, although we should not assume that a disciplining of reason entails irrationalism and it might also be argued that this is preferable to the opposing alternative, hypostazing a restricted conception of reason that abstracts itself from the life-world.

7.   James Bradley, "Chapter 34: Transformations in Speculative Philosophy, 1914-45," in *Cambridge History of Philosophy, 1870-1945*, ed. Tom Baldwin (Cambridge: Cambridge University Press, 2003), 441.

8. Michael Beaney, "Analysis," *Stanford Encyclopedia of Philosophy*, http://stanford.library.usyd.edu.au/entries/analysis/ (accessed July 30, 2010).

9. Bradley Dowden, "Time," *Internet Encyclopedia of Philosophy*, eds. Jim Fieser and Bradley Dowden, http://www.iep.utm.edu/t/time.htm (accessed July 30, 2010).

10. Bertrand Russell, *Mysticism and Logic* (London: Allen and Unwin, 1917), 21.

11. Keith Ansell-Pearson, *Philosophy and the Adventure of the Virtual* (London: Routledge, 2002), 44.

12. Ashley Woodward, "Jean-François Lyotard," *Internet Encyclopedia of Philosophy*, http://www.iep.utm.edu/l/lyotard.htm (accessed July 21, 2010); and Jean-François Lyotard, *Differend: Phases in Dispute* (Minneapolis: University of Minnesota Press, 1989). This is perhaps akin to what Miranda Fricker calls a hermeneutic injustice, where a vocabulary is not even available in which to state a given injustice. See Miranda Fricker, "Epistemic Injustice and a Role for Virtue in the Politics of Knowing," *Metaphilosophy* 34, No. 1-2 (2003): 154-73.

13. Because of its synchronic focus structuralism is difficult to accommodate within such a "temporal turn," but, as Ashley Woodward has pointed out to me, one might nonetheless resolve this problem by noting that structuralism is not, predominantly, a philosophical movement. With Derrida and the return to Heidegger, and Deleuze and a future-inflected return to Bergsonian time, we have the philosophy of time being again foregrounded.

14. Carl Honoré, *In Praise of Slow: How a Worldwide Movement Is Challenging the Cult of Speed* (London: Orion, 2004), 22.

15. Jacques Derrida, "Ousia and Gramme: A Note to a Footnote in *Being and Time*," in *Phenomenology in Perspective*, trans. Edward Casey (The Hague: Nijhoff 1970), 88-9.

16. David Allison, "Derrida and Husserl," *Understanding Derrida*, eds. Jack Reynolds and Jonathan Roffe (London: Continuum, 2004), 114.

17. Jacques Derrida, *Speech and Phenomena and Other Essays on Husserl's Theory of Signs*. trans. David Allison (Evanston, IL: Northwestern University Press, 1973), 154.

18. Antonio Negri, *Time for Revolution* (London: Continuum, 2005). Arguably this also applies to their sometime critic, Alain Badiou.

19. Emmanuel Levinas, *Time and the Other*, trans. Richard Cohen (Pittsburgh: Duquesne University Press, 1987), 76-77.

20. This summary of Deleuze's *The Logic of Sense* is indebted to James Williams who improved an earlier formulation of mine.

21. As John McCumber points out, "An assertion must be simultaneous with whatever it is that justifies it, whether that justifying factor is a state of affairs in the world or other assertions from which it follows ... The perceptual evidence for beauty of Cleopatra lies irretrievably in the past; all that can justify belief in it today is the surviving testimony." See John McCumber, *Reshaping Reason: Towards a New Philosophy* (Bloomington: Indiana University Press, 2007), 69.

22. Levinas, *Time and the Other*, 103-104.

23. See Bernard Stiegler, *Technics and Time, 1*, trans. Richard Beardsworth and George Collins (Stanford, CA: Stanford University Press, 1998); Éric Alliez, *Capital*

Times, trans. Georges Van Den Abbeele (Minneapolis: University of Minnesota Press, 1995); Jean-François Lyotard, *The Inhuman: Reflections on Time*, trans. Geoffrey Bennington and Rachel Bowlby (Palo Alto, CA: Stanford University Press, 1991).

24. Wood, *The Step Back*, 22.

25. Maurice Merleau-Ponty, "The Child's Relations with Others," in *The Primacy of Perception and other essays*, trans. W. Cobb, ed. J. Edie (Evanston, IL: Northwestern University Press, 1964), 110.

26. Bigelow thinks that presentism and four-dimensionalism are compatible, at least if relativity theories can be more modest than the usually are and not claim that they represent the whole truth, the final world on the universe. Relativity theory needs to be cleared of its positivist element and to acknowledge that acceleration, for example, is an entirely non-relative matter.

27. Robin Le Poidevin and Murray MacBeath, *The Philosophy of Time* (Oxford: Oxford University Press, 1993), 16.

28. Martin Heidegger, *Kant and the Problem of Metaphysics*, trans. R. Taft (Bloomington: Indiana University Press, 1997). Of course, it is by now a cliché that analytic philosophers tend to read the first critique while continental philosophers are more interested in the third.

29. William Newton-Smith, *The Structure of Time* (London: Routledge, 1980).

30. Huw Price, *Time's Arrow and Archimedes' Point* (Oxford: Oxford University Press, 1996).

31. Richard Swinburne, "Tensed Facts," *American Philosophical Quarterly* 27 (1990): 117-30, and Ernest Sosa, "Consciousness of the Self and of the Present," in *Agent, Language and the Structure of the World*, ed. J. Tomberlin (Bloomington, IN: Hackett, 1983), 131-43.

32. Hugh Mellor, however, deplores thought experiments that use "fantasy arguments" about time that "presume to show something possible by describing an imaginary world in which we should apparently be inclined to believe the possibility actual." For Mellor, though, showing plausibility is not enough to show possibility. See "The Unreality of Tense," in *Philosophy of Time*, eds. Robin Le Poidevin and Murray Macbeath (Oxford: OUP, 1993), 17.

33. See Dowden, "Time." See also Yuval Dolev, *Time and Realism: Metaphysical and Antimetaphysical Perspectives* (Cambridge, MA: MIT Press, 2007). Dolev suggests they both share an "ontological assumption" and intimates that a return to phenomenology might be worthwhile though he does not himself undertake this task.

34. For a good account of this, and its implications for a resultant theoretical conservatism, see James Chase, "Analytic Philosophy and Dialogic Conservatism," in *Postanalytic and Metacontinental: Crossing Philosophical Divides*, eds. Jack Reynolds, James Chase, James Williams, and Edwin Mares (London: Continuum, 2010), 85-104.

35. Nick Markosian, "Time," *Stanford Encylopedia of Philosophy*, http://stanford.library.usyd.edu.au/entries/time/ (accessed July 30, 2010).

36. Shaun Gallagher and Dan Zahavi, *The Phenomenological Mind* (London: Routledge, 2008), 19. In addition, there are some significant differences between phenomenology and poststructuralism that such a formulation would conflate. For Deleuze, transcendental philosophy and Freud allows us to get beyond time of consciousness and the time of embodied subjectivity. I become a multiplicity

of subjects living different temporalities within the same not so unified being. See Nathan Widder, "The Time Is Out of Joint—And So Are We: Deleuzean Immanence and the Fractured Self," *Philosophy Today* 50, 4 (2006): 411.

37. Dowden, "Time."

38. Gallagher and Zahavi, *Phenomenological Mind*, 40.

39. James Williams, *The Transversal Thought of Gilles Deleuze* (Manchester: Clinamen Press 2005), 114.

40. Dowden, "Time."

41. As Jim O'Shea has pointed out to me, Sellars may be one such case, given his lifelong attempts to reconcile a Kantian, experiential conception of time with the scientific image of time.

42. Körner, "Impossibility of Transcendental Deductions," 317-31.

43. Stroud, "Transcendental Arguments," 241-56.

44. David Coady reminded me that William James made this point forcefully in *The Will to Believe and Other Essays in Popular Philosophy* (New York: Dover Publications, 1956).

3. Common Sense and Philosophical Methodology: Some Metaphilosophical Reflections on Analytic Philosophy via Deleuze

1.   This chapter has benefitted from discussion with James Chase, James Williams, Jon Roffe, Ricky Sebold, Sherah Bloor, and Douglas Lackey, and was published in a different form in *Philosophical Forum* (2010).

2.   See Daniel Hutto and Matthew Ratcliffe, ed. *Folk Psychology Reassessed* (New York: Springer, 2007).

3.   And it is not simply that one can be called good and the other defective without very careful attentiveness to the environment in which this reasoning occurs, something that the analytic tradition, by and large, has not provided. See Stephen Stich, *The Fragmentation of Reason* (Cambridge, MA: MIT Press, 1993).

4.   This would be a rather narrow conception of truth, one that would seem to be committed to holding that we should not perpetuate gambler's fallacies in which we assume that several consecutive results of "heads" makes a "tail" more likely on the next throw of the coin. Empirically and psychologically speaking, the phenomenon of belief perseverance suggests that this is improbable. More philosophically, for Deleuze being preoccupied with the true/false distinction in this manner robs paradox of its intrinsic role, and tacitly posits the primacy of a good sense that is originarily oriented toward truth, whereas Deleuze's ongoing point is that this sense is socioculturally produced.

5.   Michael Friedman, *A Parting of the Ways: Carnap, Cassirer and Heidegger* (New York: Open Court, 2000).

6.   James Williams explains the Deleuzian account of this typically well in *The Transversal Thought of Gilles Deleuze*. For Deleuze, analysis presupposes the discrete nature of possibilities, denies background and connectedness of all problems to one another. Analytic philosophy's preoccupation with clear and distinct representations

deprives things of their context, ignores relationality. In particular, the association of clarity and distinctness falsely abstracts from the process of genesis and the future evolutions that are always at work in the present, interrupting it.

7. In *The Idea of Continental Philosophy*, Glendinning maintains there is no methodological, topical, or stylistic unity to continental philosophy despite its socio-political reality.

8. Friedrich Nietzsche, *On the Genealogy of Morals and Ecce Homo*, trans. Walter Kaufman (New York: Vintage, 1989), part 2 § 13, and Friedrich Nietzsche, *Will to Power*, trans. Walter Kaufman (New York: Vintage, 1968), § 409.

9. Terry Pinkard, "Analytics, Continentals, and Modern Skepticism," *Monist* 82, No. 2 (1999): 191.

10. We might also credibly assert the reverse—all philosophers are also conservative about some thing or another. Janna Thompson made these observations (separately) to me and I am inclined to agree.

11. While the moral component to this image of thought may not immediately be apparent, it is worth noting that many of the more important recent reflections on analytic methodology do, at least in my view, evince a moral element to them, either explicitly as in Jonathan Cohen's and Nicholas Rescher's linking of analytic method with the practice of democracy, or implicitly in the language and metaphors that are deployed despite the feel of neutral argument for its own sake. Cohen's *The Dialogue of Reason* (Oxford: Oxford University Press, 1986), for example, argues that there is a close connection between analytic method, democracy and non-totalitarian stances. On the other hand, one possible riposte to this is John McCumber's argument in *Time in the Ditch* (Evanston, IL: Northwestern University Press, 2001) that the rise of analytic philosophy in the United States in the McCarthy era is more than a coincidence.

12. Indeed, it amounts to something like a Holy Trinity for Bertrand Russell, who cannot forgive Plato, Spinoza, and Hegel because "they remained 'malicious' in regard to the world of science and common sense" and who criticizes Bergson because his philosophy of intuition rests on a complete condemnation of the knowledge that is derived from science and common sense. See Bertrand Russell, *Our Knowledge of the External World* (London: Routledge, 1993), 48-49.

13. Richard Rorty, *Philosophy and the Mirror of Nature* (Princeton, NJ: Princeton University Press, 1981).

14. Michael Dummett, *The Origins of Analytical Philosophy* (London: Duckworth, 1993).

15. Cohen, *The Dialogue of Reason*.

16. Frank Jackson, "Thought Experiments and Possibilities," *Analysis* 69, 1 (2009): 100-109. Of course, Timothy Williamson and others maintain that thought experiments are best seen as investigations of what's metaphysically possible, not of what's conceptually possible.

17. Russell actually is divided on this issue, making commonsense central in some places, such as *The Problems of Philosophy* (London: Oxford University Press 1912), 25, elsewhere being less committal as Søren Overgaard has pointed out to me, perhaps especially in distancing himself from ordinary language philosophy and calling the later Wittgenstein a mere lexicographer.

18. There are various methods that I cannot consider here, including the logi-

cal formalization of arguments, semantic ascent, various naturalizing techniques, etc.

19. Richard Matthews, "Heidegger and Quine on the (Ir)relevance of Logic for Philosophy," in *A House Divided: Comparing Analytic and Continental Philosophy*, ed. Carlos G. Prado (Atlantic Highlands, NJ: Humanity Books, 2003), 161.

20. David Lewis, *On the Plurality of Worlds* (London: Blackwell, 2001), 134-35. Much of the following paragraph is indebted to James Williams' analysis of this passage in *The Transversal Thought of Gilles Deleuze* (cf. 111).

21. Moreover, as Williams' observes, on Lewis' account, "conception is restricted to the concept and to properties; judgment is associated with common sense, with restricted test-cases and with pre-set logical rules; imagination is gravely restricted in terms of prior definitions of truth and consistency; and perception is associated with exact properties rather than new variations" (*Transversal Thought*, 111).

22. David Lewis, *Counterfactuals* (Cambridge, MA: Harvard University Press, 1973), 88.

23. Frank Jackson, *From Metaphysics to Ethics: A Defence of Conceptual Analysis* (Oxford: Clarendon Press, 1998), 35.

24. Nicholas Rescher, *Philosophical Reasoning* (London: Blackwell ,2001), 1.

25. Rescher, *Philosophical Reasoning*, 6, 11.

26. Max Horkheimer, *Critical Theory* (London: Continuum, 1975).

27. Rescher, *Philosophical Reasoning*, 22.

28. Rescher, *Philosophical Reasoning*, 200-201.

29. Pascal Engel, "Analytic Philosophy and Cognitive Norms," *Monist* 82, 2 (1999): 220.

30. Engel, "Cognitive Norms," 226.

31. Brian Weatherson's retort that conceptual analysis is equivalent with looking to the dictionary doesn't vitiate this circularity or offer a good argument for cognitive monism in regard to either concepts or common sense. See Weatherson, "What Good Are Counterexamples?" *Philosophical Studies* 115, (2003): 1-31.

32. Roy Sorensen, *Thought Experiments* (Oxford: Oxford University Press, 1998), 15.

33. Sorensen, *Thought Experiments*, 153. A possibility refuting use of thought experiment is rarer according to Sorensen, but it would show that one consequence of a given view might be that p does hold in a possible world, but the thought experiment can show that this is not so.

34. Deleuze's essay on Michel Tournier's rewriting of the Robinson Crusoe tale, *Friday*, *seems* to be a prolonged thought experiment (see *The Logic of Sense*). However, in this respect I think appearances are deceiving. The defining feature of thought experiments is that they are short and pithy, something that can't be said of Deleuze's essay. For Sorensen, when the thought experiment becomes a story, and one that refuses to be translated into a deductive argument, we are in the realm of literature rather than philosophy (see *Thought Experiments*, chapter 1). Another way of putting this might be to say that the Deleuzian preoccupation with Crusoe does not seem to be a consistency test or tool of any kind to sharpen distinctions. Rather, it is preoccupied with broader issues: what is our place in the world? How might we think of it otherwise? What does this show us about our normal commitments? How might we exist otherwise?

35. This latter problem is perhaps particularly acute in the debates about personal identity, which revolve around some highly imaginative scenarios. See Marguerite La Caze, *The Analytic Imaginary* (Ithaca: Cornell University Press, 2002). La Caze discusses the either/or alternatives that thought experiments in this realm usually pose, prompting us to align ourselves with either the psychological or bodily criterion of personal identity.

36. John. N. Mohanty, "Method of Imaginative Variation in Phenomenology", *Thought Experiments in Science and Philosophy*, eds. T. Horowitz and G. Massey (Lanham, MD: Rowman and Littlefield, 1991), 261-72.

37. Bernard Williams, *Utilitarianism: For and Against* (Cambridge: Cambridge University Press, 1976), 97.

38. Williams, *Transversal Thought*, 114.

39. As Husserl has shown us, lived time and the experience of a moment are produced by the retentional and protentional aspects of consciousness and cannot be understood without this.

40. Judith J. Thomson, "A Defence of Abortion," *Philosophy and Public Affairs* 1, 1 (1971): 47-66.

41. Williams, *Transversal Thought*, 138.

42. Gilles Deleuze and Felix Guattari, *What Is Philosophy?* Trans. Hugh Tomlinson and Graeme Burchill (London: Verso, 1994), 49.

43. Goodman suggests that "a rule is amended if it yields an inference we are unwilling to accept; an inference is rejected if it violates a rule we are unwilling to amend. The process of justification is the delicate one of making mutual adjustments between rules and accepted inferences; and in the agreement achieved lies the only justification needed for either." See Nelson Goodman, *Fact, Fiction, and Forecast* (Cambridge, MA: Harvard University Press, 1983), 64. While it may be a fair method to uphold in regard to epistemological justification in the philosophy of science, its importation into the realm of political philosophy (which is about more than knowledge) is more contentious.

44. See Berys Gaut, "Justifying Moral Pluralism," in *Ethical Intuitionism*, ed. Phillip Stratton-Lake (Oxford: Oxford University Press, 2003), 147, and also Brad Hooker's claim in "Intuitions and Moral Theorising" in the same journal issue (cf. 161) that most philosophers accept both the idea the method of reflective equilibrium and the idea that moral theories are better to the extent that they accord with moral claims that are attractive in their own right (i.e., intuitions), especially where they endorse a pluralist as opposed to a monist theory of value. Also see Brad Hooker, *Ideal Code, Real World* (Oxford: Oxford University Press, 2003), 10, 15. Hooker claims this also applies to moral particularists (which we might take the poststructuralists to be) because the theoretical position that there are no overarching moral or political principles that unify our various judgments must nonetheless stand or fall as a claim in relation to the diversity (or otherwise) of our intuitive commitments. In that minimal sense reflective equilibria can still be said to obtain.

45. This formulation of Rawls' position is indebted to James Chase and overlaps with our account in *Analytic Versus Continental*.

46. Thanks to Douglas Lackey for this point.

47. Thanks to James Chase for making this clear to me.

48. Carl Knight, "The Method of Reflective Equilibrium: Wide, Radical, Fal-

lible, Plausible," *Philosophical Papers* 35, 2 (2006): 222.
49. Stich, *Fragmentation of Reason*, 18.
50. Stich, *Fragmentation of Reason*, 19.
51. Transcendental arguments also tend to dogmatically assume conceptual unity. See Oskari Kuusela, "Transcendental Arguments and the Problem of Dogmatism," *International Journal of Philosophical Studies* 16, 1 (2008): 57-75. The feel of chauvinism is mitigated in some cases of transcendental argumentation, however, where the circularity in such arguments is acknowledged. See Jeff Malpas, "The Transcendental Circle," *Australasian Journal of Philosophy* 75, 1 (1997): 1-20.

4. Negotiating the Non-Negotiable: Rawls, Derrida, and the Intertwining of Political Calculation and Ultra-Politics

1. Thanks to Paul Patton and Miriam Bankovsky for comments on this chapter, which was previously published in a slightly different form in *Theory and Event* (2006).
2. Gregg Lambert, "*Une Grande Politique*, Or the New Philosophy of Right?" *Critical Horizons* 4, 2 (2003): 177-97.
3. As will become clear, Rawls' understanding of the political does not accord as much attention to the potential impact of art, for example, and it is also narrower than Derrida's on account of its preoccupation with distributive justice. In his later work, Rawls argues that political philosophy is, and should be, radically autonomous different to all other philosophical positions, something that is very different to Derrida's own account of the relationship between philosophy and politics.
4. Chantal Mouffe also makes this claim about Derrida's political significance in *The Democratic Paradox* (London: Verso, 2000), 136.
5. Following Jacques Ranciere and Etienne Balibar, Slavoj Žižek denigrates ultra-politics in various places including his essay, "Carl Schmitt in the Age of Post-Politics," in *The Challenge of Carl Schmitt*, ed. Chantal Mouffe (London: Verso, 1999), 18-37.
6. Rawls' later work will only be briefly considered in this chapter but much of the analysis contained herein arguably retains its prescience in relation to *Political Liberalism*, although I do not wish to foreclose that debate. In this text (and period), he moves from emphasizing desirability arguments (about what is rational) to feasibility arguments (what is reasonable and likely to promote stability given assumption of plurality and reasonable difference and diversity between citizens). Rawls' later work is hence no longer based in a comprehensive account of rationality (the rational calculations of the maximin principle are replaced by a priority accorded to the "reasonable" in our given historical context), and, as such, it does not so clearly offer the strict programmatic determination of the decision that is in evidence in *A Theory of Justice*. Nevertheless, certain key ideas from the earlier work persist, perhaps most notably the reliance upon the ideas of consensus and the role of intuitions in contributing to that. Moreover, he also draws an even stronger distinction between the public and political basis of justification and our more comprehensive metaphysical

private doctrines, which Derrida would problematize (although the adding up of votes might be said to be more paradigmatically "public" on Derrida's view, it is always intertwined with the "private" and they cannot be kept neatly separate, especially given the reach and saturation of the media and its mediums).

7. Ethan Macadam attempts something like this in "John Rawls at the End of Politics," *Angelaki: Journal of the Theoretical Humanities* 9, No. 3 (2004): 33-57.

8. Jacques Derrida, *Negotiations: Interventions and Interviews, 1971-2001.* ed. and trans. Elizabeth Rottenberg (Palo Alto, CA: Stanford University Press, 2002), 300.

9. For Derrida, undecidability is not the suspension of ethics and politics, as some of his critics have claimed, but the transcendental condition for any action that hopes to be called responsible, or for any decision worthy of the name. Although I think this is important, I elsewhere suggest that this transcendental condition downplays the significance of another important aspect of decision-making—the inevitability of habitual and corporeal sedimentation. Without due recognition of this, I argue that Derrida's work loses at least some of its political salience. See Jack Reynolds, "Habituality and Undecidability: A Comparison of Merleau-Ponty and Derrida on the Decision," *International Journal of Philosophical Studies* 10, 4 (2002): 449-66.

10. Derrida argues that the spectre of Marxism is, and should, continue to haunt liberal capitalism. Divested of its teleological aspects (e.g., the talk of definitive epochs in history and the exaggerated reliance on the motif of class struggle), Marxism should inspire resistance to capitalism, as well as state sovereignties, just as he insists that deconstruction does.

11. Derrida, *Negotiations*, 301.

12. Derrida, *Negotiations*, 304.

13. Derrida, *Negotiations*, 101.

14. Although I can't show this in any depth here, I think this is also the general thrust of *On Cosmopolitanism and Forgiveness* where Derrida details the problems that occur when the process of forgiving becomes a calculative program of reconciliation or amnesty that ignores the mad or undecidable aspect of it (i.e., the absolute and ethical part).

15. While the collaborative work of Deleuze and Guattari cannot be my concern here, I think that many of my comments about Derrida in this essay also apply to them. Their emphasis upon minoritarianism in *Anti-Oedipus*, for example, also explicitly contrasts itself with majoritarianism—that is, the calculative adding up of votes. Of course, Paul Patton is right to point out that their minoritarianism doesn't make them anti-democratic (contra Philippe Mengue's *Deleuze et la question de la democratie* [Paris: L'Harmattan, 2003]), because they are merely pointing out the conditions of possibility for majoritarian democratic politics (a molecularity of desire, etc.), but on my view their work retains a transcendental aspect to it that diminishes the purely calculative much as Derrida's does.

16. Karl Marx and Friedrich Engels, *The Communist Manifesto*, ed. John E. Toews (Boston: St. Martin's Press, 1999).

17. Robert Nozick, *Anarchy, State, and Utopia* (New York: Basic Books, 1974), chapter 7; and Peter Singer, "The Right to Be Rich or Poor" in *Reading Nozick: Essays on Anarchy, State, and Utopia*, ed. Jeffrey Paul (London: Basil Blackwell, 1981), 37-53.

18. As has been admitted, the extent to which Rawls' *Political Liberalism* remains committed to this kind of calculative thinking is a matter for debate. The key role that the idea of rational choice played in his early theory of justice (through the notion of the maximin principle) is no longer the focus in this latter text, and instead there is a concern with the reasonable. Nevertheless, despite this adjustment it seems to me that the Rawlsian conception of the reasonable still forecloses on what Derrida would call the incalculable because of his strict separation between substantive moral/religious questions and political questions (and between the private and the public), which Derrida would not countenance. In *The Gift of Death*, for example, the dilemma and tension between singularity and substitutability, between ethics and politics, extends beyond the Abrahamic religions to be a part of daily life: the land of Mt. Moriah that Derrida says is our habitat every second of every day. Derrida would, I think, suggest that the Rawlsian separation forecloses the possibility of genuine responsibility—which is also what Chantal Mouffe accuses Rawls of in *The Democratic Paradox*. We cannot ignore, or keep out of the public realm, the singular demands like those that God makes of Abraham (or our own loved ones make upon us) and we will return to this claim toward the end of the chapter.

19. Perhaps John Stuart Mill can be seen as an exception to this suggestion that liberalism resists consequentialism, but in my view he is better understood as a utilitarian rather than as a liberal, because his liberalism is ultimately justified in terms of a utilitarian consequentialism. Consider, for example, his famous comment that, "mankind are greater gainers by suffering each other to live as seems good to themselves, than by compelling each to live as seems good to the rest." John Stuart Mill, "On Liberty," *Three Essays* (London: Oxford University Press, 1975), 18. His controversial advocation of paternalism toward certain "primitive" indigenous societies is also thus rendered theoretically explicable.

20. Although Rawls later contends that political philosophy should be concerned with what he calls a "realistic utopianism" (see *Justice as Fairness: A Restatement* [Cambridge: Belknap Press, 2001], 4), it seems to me that in this period of his work considerations of feasibility and social stability weigh too heavily upon his purported utopianism. Similarly, the manner in which his philosophy domesticates the future (soon to be argued for) also precludes any utopianism that might constitute a genuine resistance to the present.

21. Deleuze and Guattari, *What Is Philosophy?* 99.

22. Deleuze and Guattari accord the theme of utopianism some serious attention, suggesting that utopia designates the conjunction of philosophy with the present milieu—political philosophy. They agree with Derrida that utopian political philosophy must be critical of, and resistant toward, its present time (cf. 99-100).

23. Jacques Derrida, *Points . . . Interviews, 1974-1995*, ed. Samuel Weber, trans. Peggy Kamuf (Stanford, CA: Stanford University Press, 1995). Derrida also makes a related point in "Violence and Metaphysics," in *Writing and Difference*, trans. Alan Bass (Chicago: University of Chicago Press, 1978), 100. Here Derrida distances himself from Levinas' exaltation of the wholly other in *Totality and Infinity*.

24. For two thorough analyses of auto-immunity, see Samir Haddad, "Derrida and Democracy at Risk," *Contretemps* 4 (2004): 29-38, and Samir Haddad, "Inheriting Democracy to Come," *Theory and Event* 8, 1 (2005).

25. Michael Sandel, *Liberalism and the Limits of Justice* (Cambridge: Cam-

bridge University Press, 1982), and Michael Sandel, *Democracy's Discontents* (Cambridge, MA: Harvard University Press, 1996), 3-35.

26. Mary Midgley, "Duties to Islands," in *Environmental Ethics*, ed. Robert Elliot (Oxford: Oxford University Press, 1995), 92. See also Peter Singer, *Practical Ethics* (Cambridge: Cambridge University Press, 1993), 79.

27. Honig, *Displacement of Politics*, 128.

28. Honig, *Displacement of Politics*, 141.

29. E. Rakowski, "The Future Reach of the Disembodied Will," *Politics, Philosophy and Economics* 4, 1 (2005): 91-130.

30. To be more precise, he assumes that certain foundational intuitions are shared, but suggests, as a second line of argument, that even if they are not, any divergences will play out along the lines of the main doctrines and theories he has already, in his view, refuted—that is, utilitarianism, crude intuitionism, and so on. But is this so? Are we condemned to replicate in our moral views all that has gone before, or share certain foundational intuitions? This seems suspiciously like a false dilemma.

31. Alan Brown also makes this point in *Modern Political Philosophy* (Harmondsworth: Penguin, 1986), 76. The method of reflective equilibrium is hence compatible with not just a coherentist theory of what counts as justification but also an intuitionist one, despite Rawls' attempts to obscure this point. From a critical perspective, R. M. Hare has pointed this out in *Moral Thinking* (Oxford: Clarendon Press, 1981, 75-78), as has Peter Singer in "Sidgwick and Reflective Equilibrium," *Monist* 58 (1972): 490-517.

32. Brown, *Political Philosophy*, 76-77.

33. It is also worth noting that merit is comprised not only of talent, but also hard work, which it is not so obvious should be excluded from considerations of distributive justice, as Rawls' maximin principle assumes. Will Kymlicka brings this out well through his example of a person who chooses to play social tennis for the entirety of their life and a hard-working gardener. The tennis player and the gardener had equal starting points in terms of talent and opportunity, but one chose to live a simple tennis playing life, wanting nothing more than costs for survival and tennis equipment, whereas the other earned a decent income. While Rawls argues that inequalities in income can only be justified if they improve the situation of those worst off (in this example they improve the situation of the leisured tennis player), this principle might be said to undermine equality as it entails that the hard-working gardener effectively subsidises the tennis player's lifestyle choice. Kymlicka's conclusion is that when inequalities in income are the result of choice (not talent or circumstances), Rawls' maximin principle creates, rather than removes, unfairness. See Will Kymlicka, *Contemporary Political Philosophy* (Oxford: Clarendon Press, 1990), 73-74.

34. Gilles Deleuze and Felix Guattari, *A Thousand Plateaus*, trans. Brian Massumi (Minneapolis: University of Minnesota Press, 1987), 29.

35. Hooker, *Ideal Code*, 80-81.

36. This is the residue of the old social contract tradition in his work, but, as the voluminous secondary literature on Rawls has shown us, he cannot consistently abide by this understanding of rationality. After all, as we noted earlier, from the perspective of instrumental rationality a utilitarian gamble might well be worth it and Rawls can only justify his preference for the maximin strategy under the veil of ignorance through tacitly importing conceptions of the person and, arguably, the good

life, into the equation.

37. Indeed, there are some compelling arguments for aprudentialism (understood as living for the moment spontaneously and non-deliberatively, but it might also be extended to include Derrida's future oriented insistence on madness, incalculability, etc.) that deny the maximin strategy, with all of the prudential care associated with it. See, for example, J. Trebilcot, "Aprudentialism," *American Philosophical Quarterly* 11 (1974): 203-10. For Rawls' own rejection of any such "pure time preference," cf. TJ 293.

38. Honig, *Displacement of Politics*, 131.

39. In *Political Liberalism*, Rawls continues to exclusively emphasise one aspect of political responsibility. For him, there is a special domain of the political that must remain distinct from comprehensive worldviews (especially where they involve religious convictions) and from the familial, the associational, and the personal. While individuals need to reconcile these two domains in living their lives, for Rawls this reconciliation has nothing to do with politics. While it might be interjected that surely politics requires recognition of the typical and sedimented ways in which these two domains get reconciled, Derrida's response would be that if we get rid of the aporia at the heart of politics (between singularity and substitutability), and make the personal extraneous, we end up with a formalism that makes talk of responsibility meaningless. For Derrida, we cannot exclude the apparently irrational, the radically singular, and the mad, from political life, and he would also, I suspect, disagree strongly with Rawls' dual claim that political values, in his narrow sense, normally outweigh non-political values and should continue to do so (R 139).

40. Rawls, *Political Liberalism*, 5.

41. For Derrida in *Rogues*, freedom and equality are aporetic and in conflict with one another because freedom is necessarily incalculable and equality is susceptible of calculation (R 24), but this is not to say that equality is simplistically opposed to freedom (R 48). As soon as everyone is equally free (or this is aimed for), then equality becomes interlinked with freedom and it is no longer a simple opposition between the calculability of equality and the incalculability of freedom.

42. Again, we can see teleological aspects to Rawls' work and it is worth pointing out that he is only able to envisage a state without power due to his understanding of power, which is, arguably, inadequate. In the terms of *Political Liberalism*, power is understood as repressive and "coercive" (cf. 139-40), rather than productive and omnipresent in the Foucauldian sense.

43. William Connolly, *Politics of Ambiguity* (Wisconsin: University of Wisconsin Press, 1987), 10.

44. In *Political Liberalism*, Rawls argues for the possibility (and likelihood) of an overlapping consensus between an individual's disparate comprehensive moral worldviews and the public and political conception of justice that he proposes (cf. 134). For Rawls, our private worldviews and our considered and public political judgments, although capable of being very different such as to permit different religious views, are nonetheless fundamentally compatible with each other, not just within ourselves but also with the political views of others.

45. Rawls suggests that the principle of the law of nations is a principle of equality (TJ 378) and he certainly does not accept that all states are rogue states as Derrida suggests. Indeed, Derrida argues for the establishment of a 'New International'

because, on his view, the law of nations (sovereignty) relies on the principle of might is right, or the reason of the strongest, as he suggests in the chapter of the same name in *Rogues*.

5. The Politics of Futurity in Derrida and Deleuze

1. This chapter was first delivered at a conference at the Australian National University entitled "Unassumable Responsibility," and was subsequently published in *Borderlands* (2004).

2. It is sometimes suggested that Deleuze and Guattari's characterization of the post-Husserlian philosophers as committed to instituting "transcendence in immanence" is *What Is Philosophy?* (46) is intended to constitute a dig at Derrida, but there is little textual support for this. On the other hand, Derrida does seem to directly criticize Deleuze and Guattari's collaborative works in *On Touching* (cf. 125) for remaining within a certain haptocentric tradition and we will consider his reservations about their idea of philosophy as concept creation shortly

3. Gordon Bearn, "Differentiating Derrida and Deleuze," *Continental Philosophy Review* 33 (2000): 467.

4. Daniel W. Smith, "Deleuze and Derrida, Immanence and Transcendence," in *Between Derrida and Deleuze*, eds. Paul Patton and John Protevi (London: Continuum, 2003), 46.

5. Jacques Derrida, "Force of Law," *Deconstruction and the Possibility of Justice*, ed. and trans. Drucilla Cornell (New York: Routledge, 1992), 25.

6. John Caputo, *Deconstruction in a Nutshell* (New York: Fordham University Press, 1997), 136.

7. Derrida, "Force of Law," 26.

8. Derrida, *Negotiations*, 11-40.

9. See John Caputo, *The Prayers and Tears of Jacques Derrida* (Bloomington: Indiana University Press, 1997); and Kevin Hart, *The Trespass of the Sign: Deconstruction, Theology and Philosophy* (Cambridge: Cambridge University Press, 1989), as well as the various works of Jean-Luc Marion.

10. Derrida, as quoted in Caputo, *Deconstruction in a Nutshell*, 24.

11. Jacques Derrida, *Psyche: Inventions of the Other, 1*, eds. Peggy Kamuf and Elizabeth Rottenberg (Palo Alto, CA: Stanford University Press, 2007), 45.

12. I refer here to Deleuze's ontological thesis of the univocity of being, of which, in *Difference and Repetition*, he details three key historical stages: that of Duns Scotus, Spinoza, and finally Nietzsche, whose thought of the eternal return of difference is said to finally accomplish this univocity, this ontology of immanence (cf. 39-41).

13. James Williams, *A Critical Introduction and Guide to Gilles Deleuze's Difference and Repetition* (Edinburgh: Edinburgh University Press, 2003), 206.

14. Jon Roffe, "Deleuze," *Internet Encyclopedia of Philosophy*, eds. James Fieser and Brad Dowden, http://www.iep.utm.edu/deleuze.htm (accessed March 2009). For more on the philosophy of time and chapter 2 of *Difference and Repetition*, see James Williams, *Gilles Deleuze's Philosophy of Time* (Edinburgh: Edinburgh University Press 2011).

15. Friedrich Nietzsche, *The Gay Science*, ed. Bernard Williams, trans. Josefine Nauckhoff and Adrian Del Caro (Cambridge: Cambridge University Press, 2001), 341.

16. Gilles Deleuze, *Pure Immanence: Essays on A Life*, trans. Anne Boyman (New York: Zone Books, 2001), 88. This text also discusses the eternal return of difference.

17. Deleuze and Guattari, *A Thousand Plateaus*, 198-99.

18. See Paul Patton, *Deleuze and the Political* (London: Routledge, 1999); Brian Massumi, *A User's Guide to Capitalism and Schizophrenia: Deviations from Deleuze and Guattari* (Cambridge, MA: MIT Press, 1992); and Eugene Holland, *Deleuze and Guattari's Anti-Oedipus: An Introduction to Schizoanalysis* (London: Routledge, 1999).

19. Derrida, *Psyche*, 102.

20. See John Protevi and Paul Patton, ed. *Between Deleuze and Derrida* (London: Continuum, 2003).

6. Wounds and Scars: Deleuze on the Time and the Ethics of the Event

1. This chapter was initially published in *Deleuze Studies* (2007). In the next issue of the same journal, James Williams disputes some of my central claims and I reply.

2. G. W. F. Hegel, *Phenomenology of Spirit*, trans. A. V. Miller, J. N. Findlay (Oxford: Oxford University Press, 1979), 407.

3. In the context of his consideration of the Stoic response to catastrophic events, Deleuze consistently treats the event and the wound as synonyms. While his student, Éric Alliez, hence traces a divergence between pensée 68 and the phenomenologists on the issue of their particular understanding of the event and the rupture that the former claim it induces with phenomenological intentionality, this essay proceeds by pointing to an isomorphic and almost synonymous aspect: the wound. The event is wounding for pensée 68 (although Deleuze is my focus here, elsewhere I extend this analysis to the thought of Derrida). Unlike Alliez, however, I do not unreservedly endorse this turn. See Éric Alliez, "Questionnaire on Deleuze," *Theory, Culture and Society* 14, 2 (1997): 82.

4. Elizabeth Grosz, *The Nick of Time* (Sydney: Allen and Unwin, 2004).

5. Jack Reynolds and Jon Roffe, "Deleuze and Merleau-Ponty: Immanence, Univocity, and Phenomenology," *Journal of the British Society of Phenomenology* 37, 3 (2006): 228-51.

6. Deleuze discusses the virtual and the actual in more detail in the essay of that name included in *Essays Critical and Clinical*, trans. Daniel W. Smith (Minneapolis: University of Minnesota Press, 1997) and in *Difference and Repetition, What Is Philosophy?* and *Bergsonism*.

7. See Negri, *Time for Revolution*. A very different account of Deleuze's work, and the significance of this independence from matter, can be found in Peter Hallward's provocative but often insightful book, *Out of This World: Deleuze and the*

Philosophy of Creation (London: Verso, 2006).

8.	Despite what is arguably a more complicated understanding of this purported "independence," Derrida argues for a position that is actually surprisingly closely related: in his work, habit, the present, states of affairs (the realm of the possible and the actual), are all tacitly marginalized in contrast to the "impossible" or unconditional aspect of the various aporias that preoccupy him. Moreover, against Daniel W. Smith's analysis of Derrida's talk of impossibility as being but a form of *ressentiment* and the raising of impotency to a value, it seems to me that the work of Deleuze and Derrida bear more in common than is usually thought. If there is an extra-worldly ethic in Deleuze's work, a temporal and transcendental impetus given to that which is not, just as there is in Derrida's work, then perhaps the parallels between these two French philosophers are worthy of more sustained consideration. For Smith's account of their relation, see Daniel W. Smith, "Deleuze and the Question of Desire: Towards an Immanent Theory of Ethics," *Parrhesia* 2 (2007): 69; and Daniel W. Smith, "Deleuze and Derrida."

9.	James Williams' first chapter in *The Transversal Thought of Gilles Deleuze* poses this problem well.

10.	In *Out of This World*, Peter Hallward suggests that they ultimately conflate into a monistic univocity, precisely because what I label the transcendental component of the distinction, that which does not refer to *l'actualite*, is in fact ultimately all that there is. Deleuze, however, consistently speaks of a "secret dualism" in *The Logic of Sense*, with, as we have seen, the body and states of affairs the lesser but arguably not entirely effaced term of the dualism.

11.	Alain Badiou, *Deleuze: The Clamour of Being*, trans. Louise Burchill (Minneapolis: University of Minnesota Press, 2000). Also, see Jon Roffe's translation of a chapter from Badiou's *Logiques des Mondes* (Paris: Editions de Seuil, 2006), titled "The Event in Deleuze," *Parrhesia* 2 (2007): 38.

12.	Deleuze complicates this account towards the end of *The Logic of Sense* in the series titled "Aion," where he details the different modalities of the present and instant that are characteristic of Aion and Chronos. Adequately addressing this material, which is in tension with some of the other formulations in this book, is beyond me here.

13.	Hallward, *Out of This World*.

14.	Williams, *Transversal Thought*, 16-17. Although Williams poses this question, he ultimately would not agree with my attempts to problematize Deleuze's position here.

15.	Again, see Smith, "Deleuze and the Question of Desire." This admirably clear and precise essay encapsulates a certain thrust of Deleuze's ethics, but it has relatively little to say about the extra-worldly ethic of *The Logic of Sense*, despite the fact that Smith is cited on the back of Hallward's *Out of This World*, which devotes considerable time to explicating this extra-worldly ethic.

16.	Mark Hansen, "Becoming as Creative Involution? Contextualising Deleuze and Guattari's Biophilosophy," *Postmodern Culture* 11, 1 (2000).

17.	John Sellars, "An Ethics of the Event: Deleuze's Stoicism," *Angelaki: Journal of the Theoretical Humanities* 11, 3 (2006): 164.

18.	Sellars, "Deleuze's Stoicism," 161.

19.	James Williams, "Deleuze and J. M. W. Turner: Catastrophism in Philosophy," in *Deleuze and Philosophy*, ed. Keith Ansell-Pearson (London: Routledge, 1997), 232-46. For a rather different interpretation, see Hallward, *Out of This World*.

20. James Williams, "Deleuze and Whitehead: The Concept of Reciprocal Determination" in *Deleuze, Whitehead and the Transformation of Metaphysics*, eds. A. Cloots and K. Robinson (Brussels: Konklijke Vlaamse Academie Van Belgie Voor Wetenschaapen En Kusten, 2005), 89-105.

21. Philippe Mengue, *Deleuze et la question de la démocratie* (Paris: L'Harmattan, 2003).

22. In this respect, it is also worth considering the central role that time plays in Merleau-Ponty's famous account of ambiguity in *Phenomenology of Perception*, and the manner in which it breaches what has been considered to be inner and outer. The philosophy of ambiguity is a philosophy of the scar.

23. See Shaun Gallagher, "The Place of Phronesis in Postmodern Hermeneutics," *Philosophy Today* 37 (1993): 298-305. This paper makes it apparent that phenomenological phronesis is generally rejected in postmodern positions (as in Derrida and Deleuze, where it is understood to be common sense), or inadequately thematized, as Gallagher suggests is the case with Lyotard. It seems to me that Gallagher is right that there is a tacit denial of the importance of phronesis (or Gadamerian *Verstehen*), habits, ethos over time, and that these are replaced by a priority given to inventing new moves, new games, particularly in the case of Lyotard and Deleuze. While they know that creating of new games is never ex nihilo, they assert there is a temporal priority given to a particular futural synthesis of time that for them is exemplified by motifs like the dice throw, the child, etc.

24. Deleuze and Guattari, *What Is Philosophy?* 160, 159.

7. Deleuze on the "Perverse" Structure: Beyond the Other-Structure and the Struggle for Recognition

1. This chapter selectively synthesises a number of different publications of mine on Deleuze, including *Parrhesia* (2006), *Symposium* (2008), and *Angelaki* (2010).

2. With the help of a sustained reading of Spinoza and Marx, Hardt and Negri transform Deleuze's concept of a multiplicity into the more clearly political idea of the multitude. See Michael Hardt and Antonio Negri, *Empire* (Harvard, MA: Harvard University Press, 2000).

3. Gilles Deleuze and Félix Guattari, *Anti-Oedipus: Capitalism and Schizophrenia*, trans. Robert Hurley, Mark Seem and Helen Lane (New York: Viking Press, 1977), 100.

4. James Williams explores some of the differences between the thought of Levinas and Deleuze well in chapter 3 of *The Transversal Thought of Gilles Deleuze*. Nevertheless, his positive interpretation of Deleuze's view of the other-structure, including the association of it with the virtual and the transcendental (cf. 36, 39), will be challenged in the remainder of this chapter by pointing to passages that trouble Williams' interpretation. Deleuze consistently treats the other-structure ambivalently,

whether it be in regard to his enigmatic comments about going beyond the other-structure in *Difference and Repetition*, as well as the entire essay on Michel Tournier's novel, which explores the priority of a certain perverse-structure (which is linked to what Deleuze calls the virtual/transcendental) over the other-structure (which is treated more like an actualization of the transcendental).

5. Deleuze and Guattari, *What Is Philosophy?* 46.

6. Maurice Merleau-Ponty, *The Visible and the Invisible*, trans. Alphonso Lingis (Evanston, IL: Northwestern University Press, 1964), 79. While Sartre asserts that the alienating experience of being-for-others (e.g., shame before the look) precedes and founds our experience of being-with-others as a collective group, Merleau-Ponty argues that being-with-others, not being-for-others, is the more primordial mode. Although Deleuze would have reservations about the connotations of sharing involved in this "being-with," he seems committed to agreeing with Merleau-Ponty that the phenomenological expressivity of the embodied other comes first, before the conflictual relations evinced by the look. For more on this, see Reynolds and Roffe, "Merleau-Ponty and Deleuze".

7. Deleuze and Guattari, *What is Philosophy?* 17-9.

8. It hence appears that Deleuze defines the other-structure as against resemblance and similarity, and in favor of heterogeneity and difference. It is for this reason that Williams argues that Deleuze's conception of the other as "structure of the possible" can be understood as referring to the virtual, rather than merely to the actual or to the possible in the derogatory sense that he usually gives to this term (Williams, *Transversal Thought*, 36, 39). Although this interpretation seems plausible at first glance, it does not quite capture the ambivalence of Deleuze's writings on the issue, including his references to a "beyond" of the other-structure (DR 282), as well as his explicit comments that treat the other-structure as of the order of possibility and resemblance (LS 345-6), as we will see. Other literature on this issue includes: Peter Hallward, "Deleuze and the world of others," *Philosophy Today* 41, 1 (1997): 530-44; Sean Bowden, "Deleuze et les Stoïciens: une logique of l'événement," *Bulletin de la Societe Américaine de Philosophie de Langue Française* 15, 1 (2005): 72-97; S. Arnott, "The Problem of Solipsism and Deleuze's Ethics," *Contretemps* 2 (2001); and the two essays by Constantin Boundas that are discussed in n12 below.

9. Jean-Paul Sartre, *"No Exit" and Three Other Plays*, trans. S. Gilbert and L. Abel (New York: Vintage, 1956).

10. Williams, *Deleuze's 'Difference and Repetition'*. Williams reads these as the two dominant motifs of Deleuze's book.

11. Williams, *Transversal Thought*, 36, 39.

12. Boundas suggests that "the foreclosure of the other discloses a world of necessity, where the virtual and the possible can no more find firm foothold" (111), and "the absence of the other as possible world would bring about the collapse of the possible and the triumph of the necessary" (112). This conclusion doesn't seem quite right for the opposite reasons to Williams' account: although Deleuze is ambiguous on this, the only coherent interpretation is that the foreclosure of the other-structure in favor of the perverse-structure *does* give us the order of the virtual (and the time of Aion). Indeed, at one point Boundas himself admits that the foreclosure of the other-structure "allows, for the first time, for the emergence of the virtual," and it is no easy task to see how this claim might be reconciled with his earlier comments. For

Boundas' own analysis, see "Deleuze: Serialisation and Subject Formation," *Deleuze and the Theatre of Philosophy*, eds. Constantin Boundas and Dorothea Olkowski (New York: Routledge, 1994). He offers a more detailed account of this interpretation in Constantin Boundas, "Foreclosure of the Other: From Sartre to Deleuze," *Journal of the British Society of Phenomenology* 24, 1 (1993): 32-41.

13. For Deleuze, the discovery of the surface and the critique of depth is a theme of modern literature in general, not just Lewis Carroll's work which is his focus in *The Logic of Sense*, and it is something that he also seems to endorse, notwithstanding his love for Artaud, who is, for Deleuze, a writer of the depths.

14. Rosalyn Diprose, *Corporeal Generosity: On Giving with Nietzsche, Merleau-Ponty and Levinas* (Albany, NY: SUNY Press, 2002).

15. Michel Foucault, "Theatrum Philosophicum," in *Language, Counter-Memory, Practice*, ed. Donald F. Bouchard (Ithaca, NY: Cornell University Press, 1977), 170.

16. It is interesting to note that Crusoe's three-stage journey in relation to the other-structure parallels what Deleuze describes as the three-stage journey of Oedipus and Hamlet in *Difference and Repetition* and which he argues was also projected to be the case with Nietzsche's Zarathustra—i.e., from sickness, to convalescence, to the overcoming that he associates with the future and the eternal return of difference (DR 89-92, 298). In all cases, in attaining to the final stage they manage to live the transcendental and the virtual and thus also function as a kind of moral exemplar, either of the play of surfaces and the time of Aion (Crusoe in *The Logic of Sense*), or of the differential time of the future (Zarathustra, Oedipus, and Hamlet in *Difference and Repetition*).

17. Deleuze and Guattari, *What Is Philosophy?* 10, 108.

18. Deleuze and Guattari, *What Is Philosophy?* 151.

19. Gilles Deleuze, "Coldness and Cruelty," *Masochism*, trans. Jean McNeil (New York: Zone Books, 1991), 40.

20. More accurately, Sartre sees no possibility of redemption in *Being and Nothingness*. His abandoned but subsequently published *Notebooks for an Ethics* (trans. David Pellauer, Chicago: University of Chicago Press, 1992) are rather more optimistic.

21. Jean-Paul Sartre, *Being and Nothingness*, trans. Hazel Barnes (London: Routledge, 1994), 259.

22. Michele Le Doeuff, *Hipparchia's Choice: An Essay Concerning Women, Philosophy etc*, trans. T. Selous (Oxford: Blackwell, 1991), 62-63.

23. Sartre, *Being and Nothingness*, 377.

24. Deleuze, "Coldness and Cruelty," 72.

25. Deleuze, "Coldness and Cruelty," 71.

26. Gilles Deleuze, "From Sacher-Masoch to Masochism," trans. Christian Kerslake, *Angelaki: Journal of the Theoretical Humanities* 9, 1 (2004): 126.

27. It is well-known that Freud's metapsychological model changed throughout his career, but less recognized is the transformation of his position on sadism and masochism which included reversing the order of priority that he thought obtained between them. In his famous 1905 essay, "Three Essays on Sexuality," sadism was classified as one of the component instincts of sexuality, with masochism a secondary phenomena, an inversion of sadism. Later on in his work, however, he argued for a

rather different distinction: one between life instincts and death instincts, and it was the phenomena of sadism and masochism which led to this later hypothesis and the famous positing of a death instinct in *Beyond the Pleasure Principle* and elsewhere. At this stage, he also maintained that masochism or internal cruelty was more fundamental than aggression against others. He didn't completely recapitulate his earlier view in that he argued for a death instinct rather than an aggressivity instinct, but he certainly argued for the priority of the death instinct through the complex histories of moral masochism that seemed to him to point to a temporal priority of inward aggression over outer aggression.

28. Jessica Benjamin, *The Bonds of Love* (New York: Pantheon Books, 1988), 56.

29. This is the claim proposed by Nathan Widder with which I agree. Widder, "Time Is Out of Joint," 413.

8. Derrida, Friendship, and the Transcendental Priority of the "Untimely"

1. An earlier version of this chapter was published in *Philosophy and Social Criticism* (2010).

2. See, for example, *Of Hospitality*, trans. Rachel Bowlby (Palo Alto, CA: Stanford University Press, 2000); *Given Time*, trans. Peggy Kamuf (Chicago: University of Chicago Press, 1992); and *On Cosmopolitanism and Forgiveness*, trans. Mark Dooley and Joe Hughes (London: Routledge 2001).

3. Deleuze describes the quasi-cause as the manner in which the "virtual" haunts and at least partly produces the actual. Quasi-causality, however, does not function on the basis of strict causal necessitation and determination, but abides by a logic of expression.

4. Derrida's reference to the virtual in this context seems to explicitly and deliberately invoke connections with the work of Deleuze. I think this happens more and more frequently in Derrida's texts, where references to Deleuzian understandings of good and common sense frequently recur. This is not to deny that *On Touching* is critical about Deleuze and Guattari's work in a number of places, most notably for the manner in which Derrida suggests that it ultimately perpetuates a certain haptocentric intuitionism (with concepts like the body-without-organs), and thus remains at least partly tied to the Christian onto-theological tradition (cf. 125).

5. Likewise Aristotle's tacit binary opposition between loving and being loved, and the suggestion that the former is on the side of life and the latter on the side of death, is also convincingly deconstructed. I cannot address this in any detail here, however.

6. Jacques Derrida, *Adieu to Emmanuel Levinas*, trans. Pascale-Anne Brault and Michael Naas (Palo Alto, CA: Stanford University Press, 1999); Derrida, *The Work of Mourning*, eds. and trans. Pascale-Anne Brault and Michael Naas (Chicago: University of Chicago Press, 2001); Derrida, *Memoires: for Paul de Man*, trans. N. Lindsay, J. Culler, E. Cadava, and P. Kamuf (New York: Columbia University Press, 1989).

7. Derrida sometimes refers to this mourning as originary, sometimes as pre-originary: in *Aporias* it is originary, perhaps deliberately so on account of his engagement with Heidegger, whereas in *On Touching* it is pre-originary.

8. Jacques Derrida, "Otobiographies" in *Ear of the Other* (London: University of Nebraska Press, 1985); Jacques Derrida *Negotiations: Interventions and Interviews, 1971–2001*, ed. and trans. Elizabeth Rottenberg (Palo Alto: Stanford University Press, 2002), 46-54.

9. A good discussion of Derrida's arguments in this respect can be found in Richard Beardsworth, "Jacques Derrida: The Power of Reason," *Theory and Event* 8, 1 (2005).

10. To provide but two examples, consider *Negotiations* (cf. 231), and *Demeure: Fiction and Testimony*, trans. Elizabeth Rottenberg (Palo Alto, CA: Stanford University Press, 2000), 16.

11. In *Gift of Death*, for example, genuine responsibility consists in oscillating between the demands of that which is wholly other (in Abraham's case, God, but also any particular other) and the more general demands of a community, and in enduring this trial of the undecidable decision rather than simply resolving it (cf. 70).

12. At the same time, Derrida's own descriptions of pre-originary mourning must also contain some phenomenological or psychoanalytic register.

13. Jacques Derrida, *A Taste for the Secret*, trans. G. Donis (London: Polity Press, 2001), 40. My italics.

14. See Jack Reynolds, *Understanding Existentialism* (Chesham, UK: Acumen, 2006). The final chapter of this book makes some similar observations on the relationship that obtains between existentialism and Derrida's work.

15. It is omni-temporal rather than presentist, even according to Derrida's own account. Moreover, for Merleau-Ponty the habitual action is not based merely in a temporality of the present, and yet nor is it restricted only to the past. The presence of habituality is built upon our past-learned skill that is still in play, and which, nevertheless, must also open us to slightly different and unanticipated scenarios. So even the mode of existence in which we unthinkingly react partakes in a previous existence that has engendered certain results. This is what allows us to anticipate eventual outcomes, and yet it also necessitates precipitation and the hastening of a coming event, and these two aspects mutually encroach such that we condition and alter the world, just as the world also conditions and produces us. The apparent presence involved in behaving habitually is hence always internally divergent, requiring both anticipative and precipitative elements that never resolve themselves into any absolute stability that might be denigrated as conforming to the metaphysics of presence.

9. Time Out of Joint: Between Phenomenology and Poststructuralism

1. An earlier version of this chapter was given as a keynote address at the conference, "Time, Transcendence and Performance," at Monash University, Australia, and subsequently published in *Parrhesia* (2010) in the context of a critical engage-

ment with the work of Nathan Widder.

2. Edmund Husserl, *Ideas 1*, trans. F. Kersten (The Hague: Nijhoff 1982), §24.

3. Derrida, *Speech and Phenomena*, 62-3.

4. Nathan Widder, *Reflections on Time and Politics* (Pennsylvania, PA: Pennsylvania State University Press, 2008).

5. Einstein contends, for example, that when it comes to time there is just the objective time revealed by physics, and psychological or subjective time on the other hand, and nothing else to be said about time. See Henri Bergson, *Duration and Simultaneity* (Manchester: Clinamen Press, 1999), 159. Suffice to say almost no continental philosopher will accept that.

6. Widder, *Time and Politics*, 4; and James Williams, "Why Deleuze Does Not Blow the Actual on Virtual Priority. A Rejoinder to Jack Reynolds," *Deleuze Studies* 2, 1 (2008): 97-100.

7. As well as the more explicit focus on place of his later work, this is, I think, a key reason behind Jeff Malpas preferring the later Heidegger to the thinker of *Being and Time*—see Malpas, *Heidegger's Topology* (Cambridge, MA: MIT Press, 2007). Charles Taylor also makes some related points in "The Validity of Transcendental Arguments," *Philosophical Arguments* (Cambridge, MA: Harvard University Press, 1995), 20-33.

8. William Blattner, *Heidegger's Temporal Idealism* (Cambridge: Cambridge University Press, 1999), 130.

9. Blattner, *Heidegger's Idealism*, 149, 151.

10. Shaun Gallagher, *How the Body Shapes the Mind* (Oxford: Oxford University Press, 2005), 78. Also see Gallagher and Zahavi, *The Phenomenological Mind*, 206.

11. Blattner, *Heidegger's Idealism*, 170-71.

12. Sartre, *Being and Nothingness*, 487.

13. John Sutton, "Batting, Habit, and Memory: The Embodied Mind and the Nature of Skill," *Sport in Society* 10, 5 (2007): 763-86.

14. Edmund Husserl, "On the Phenomenology of the Consciousness of Internal Time," *Collected Works*, ed. R. Bernet, trans. J. Brough, Volume 4 (Dordrecht: Kluwer, 1991), 23-31.

15. Sutton, "Batting, Habit and Memory."

16. Hubert Dreyfus, *What Computers Still Can't Do: A Critique of Artificial Reason* (Cambridge, MA: MIT Press, 1997). Of particular relevance is part 3, "The Role of the Body in Intelligent Behaviour."

17. As such, phenomenology of this kind is not greatly trouble by Benjamin Libet's experiments on time-consciousness that seem to show that our conscious experience of making decisions is actually misleading. When we are conscious of having made a decision, in actual fact the decision was made (judging by neural activity) about 300 milliseconds earlier. What do such findings mean for phenomenology? Such data would be taken by most phenomenologists to support (rather than falsify) their view, in that embodied intentionality is shown to operate at a different level from conscious reflective decision-making (roughly the know-how/know-that distinction), and the manner in which the former kind of prereflective motor intentionality is always-already at work.

18. Hoy, *The Time of Our Lives*, 51.

19. Widder, "Time Is Out of Joint," 411.

20. Heidegger says "the attempt in *Being and Time*, §70, to derive human spatiality from temporality is untenable." See Martin Heidegger, *On Time and Being*. trans. Joan Stambaugh (New York: Harper, 1972). Also, see Merleau-Ponty, *Phenomenology of Perception*, 482.

21. Friedrich Nietzsche, *On the Genealogy of Morals and Ecce Homo*, trans. Walter Kaufman (New York: Vintage Books, 1969), 57-58.

22. See Derrida's *Spectres of Marx* for an interesting account of the differing French and German translations of this famous quote.

23. Derrida, *Spectres*, 77.

24. Leonard Lawlor, "Derrida," *Stanford Encyclopedia of Philosophy*, http://plato.stanford.edu/entries/derrida/ (accessed June 12 2010).

25. Derrida, *Spectres*, 109.

26. Widder, *Time and Politics*, 3, 6.

27. Deleuze and Guattari, *A Thousand Plateaus*, 418.

10. Dreyfus, Merleau-Ponty, and Deleuze on *L'Habitude*, Coping, and Trauma in Skill Acquisition

1. An earlier version of this chapter was published in *International Journal of Philosophical Studies* (2006).

2. And questions of home can never be wholly disassociated from questions of habit and inhabiting—for more on this, see Scott Weiner's "Inhabiting in the *Phenomenology of Perception*," *Philosophy Today* 34, 4 (1990): 342-53, and David Morris, *The Sense of Space* (Albany, NY: State University of New York Press, 2004).

3. Although Dreyfus' general exegetical position seems to me to be correct, there are, of course, other possible interpretations of Merleau-Ponty and Heidegger. We might also note that Bruin Christensen challenges Dreyfus' anti-representationalist reading. See, for example, Carleton B. Chistensen, "Getting Heidegger off the West Coast," *Inquiry* 41, 1 (1998): 65-87. In relation to Merleau-Ponty, aspects of Dreyfus' interpretation have also been contested by Sara Heinamaa and Komarine Romdenh-Romluc, among others.

4. Hubert Dreyfus and Stuart Dreyfus, "The Challenge of Merleau-Ponty's Phenomenology of Embodiment for Cognitive Science," in *Perspectives on Embodiment: The Intersections of Nature and Culture*, eds. Homi Haber and Gail Weiss (London: Routledge, 1999): 109-10.

5. Dreyfus, *What Computers Can't Do*. '

6. Caroline Davies, "As the Crow Flies: Follow the A34 and Turn Right," *The Age*, reprinted from the *Telegraph*.

7. Hubert and Stuart Dreyfus, "The Challenge of Merleau-Ponty's Phenomenology," 103-20. For a more detailed treatment of the phenomenology of skill acquisition, see Hubert Dreyfus and Stuart Dreyfus, *Mind Over Machine* (New York: Free Press, 1982).

8. Gareth Evans, as cited in Sean Kelly, "Merleau-Ponty and the Body," in *The*

Philosophy of the Body, ed. Michael Proudfoot (London: Blackwell, 2001), 62-76.

9. Martin Dillon, *Merleau-Ponty's Ontology* (Bloomington: Indiana University Press, 1988), 138.

10. Sean Kelly, "Grasping at Straws: Motor Intentionality and the Cognitive Science of Skilled Behaviour," in *Essays in Honour of Hubert L. Dreyfus, Volume 2: Heidegger, Coping and Cognitive Science*, eds. Mark Wrathall and Jeff Malpas (Cambridge, MA: MIT Press, 2000), 161-78.

11. Kelly, "Grasping at Straws," 167.

12. Hubert Dreyfus and Stuart Dreyfus, "The Ethical Implications of the Five-Stage Skill-Acquisition Model," *Bulletin of Science, Technology, and Society* 24 (2004): 251-74. See also Hubert Dreyfus' website and the essay, "What Is Moral Maturity? A Phenomenological Account of the Development of Ethical Expertise," http://www.nuc.berkeley.edu/courses/classes/E-124/Moral_Maturity_8_90.pdf

13. Patton, *Deleuze and the Political*, 75.

14. Of course, the poststructuralists are also concerned with the genesis and preconditions of subjectivity and individuation. Both Deleuze's transcendental empiricism and Foucault's genealogy are heavily invested in the role of the past in the constitution of subjectivity, but never in quite the same way as the communitarians.

15. See Merleau-Ponty's *Humanism and Terror: An Essay on the Communist Problem*, trans. John O'Neill (Boston: Beacon Press, 1969); and de Beauvoir's *The Ethics of Ambiguity*, trans. Bernard Frechtman (New York: Kensington Publishing, 1976). It is only the persistence in their work of a version of the master-slave dialectic that precludes their political philosophy from being a form of communitarianism.

16. Axel Honneth reaffirms this connection between phenomenology and forms of communitarianism when he suggests that Michael Sandel's famous critique of the neutral subject of Rawls' *A Theory of Justice* tacitly relies upon a phenomenology. See Axel Honneth, *The Fragmented World of the Social*, ed. Charles. W. Wright (Albany, NY: SUNY Press, 1995), 234. Although it might not need repeating, Sandel argues that Rawls' model for determining distributive justice—self-interested calculations behind a "veil of ignorance"—presupposes an isolated and autonomous self that is capable of being divorced from the ends that it has, as well as the goals of the society that it is a part of. While there are ways of rescuing Rawls from this criticism as we saw in chapter 4, Sandel opposes this Rawlsian subject to the thoroughly situated subject that his own unacknowledged phenomenology reveals to us.

17. Dreyfus, "What is Moral Maturity?" 2.

18. Even if we disagree with Sandel's critique, it nevertheless remains the case that the abstract and self-interested rationality involved in Rawls' hypothetical veil of ignorance provides us with a rough guide as to the fairness or otherwise of our empirical choices, by forcing us to ignore factors that are peculiar to our own situation and station in life. As such, Rawls begins with rationality, rather than with what happens in our everyday ethical coping.

19. See John Stuhr's analysis of Deleuze's link with the American pragmatists in *Pragmatism, Postmodernism, and the Future of Philosophy* (London: Routledge, 1993).

20. This relation between phenomenology and what Deleuze denigrates as ur-doxa is explored at length in Reynolds and Roffe, "Merleau-Ponty and Deleuze." Other literature that addresses the relation between Deleuze and Merleu-Ponty in-

cludes: Len Lawlor, "The End of Phenomenology: Expressionism in Deleuze and Merleau-Ponty,", *Continental Philosophy Review* 31 (1998): 15-34; Gail Weiss, "Écart: The Space of Corporeal Difference," in *Chiasms: Merleau-Ponty's Notion of Flesh*, eds. Fred Evans and Len Lawlor (Albany, NY: SUNY Press, 1995); and D. Taylor, "Phantasmatic Genealogy" in *Merleau-Ponty, Hermeneutics and Postmodernism*, eds. Thomas Busch and Shaun Gallagher (Albany, NY: SUNY Press, 1992).

21. For Deleuze, habit is said to underlie the organization of the Id and he also insists that the problem of habit is badly understood if it is subordinated to pleasure (Merleau-Ponty agrees, and this is why he argues that his notion of equilibrium is preferable to traditional understandings of behavior as goal-oriented, motivated by pleasure, etc.). In fact, Deleuze goes as far as to say: "habit precedes the pleasure principle and renders it possible" (DR 97). While this does not entail the subsumption of the unconscious under the umbrella of "habituality," it is important to also note the formulation that Deleuze consistently repeats: "I do not repeat because I repress. I repress because I repeat" (DR 18). Taken together, these enigmatic comments seems to suggest that the unconscious is part of repetition and habit, rather than something distinct and separable from them as the Freudian metapsychological model tends to imply. A more detailed analysis of this relation can be found in Keith Faulkner, *Deleuze and the Three Syntheses of Time* (New York: Peter Lang, 2005) and Christian Kerslake, *Deleuze and the Unconscious* (London: Continuum, 2007).

22. Charles Spinosa, Fernando Flores and Hubert Dreyfus, *Disclosing New Worlds: Entrepreneurship, Democratic Action, and the Cultivation of Solidarity* (Cambridge, MA: MIT Press, 1999).

23. Patton, *Deleuze and the Political*, 46.

24. This trace of an old-fashioned empiricism is apparent in other places in *Difference and Repetition*, despite Deleuze's avowed transcendental empiricism, which insists on reciprocal determination between what he calls the virtual (a transcendental realm of differences that sunders identities) and the actual (which includes all transient identities, including habitual one's). At one stage, for example, Deleuze concludes: "in other words there is no ideo-motivity, only sensory-motivity" (DR 23). Although this might seem to resonate with Merleau-Ponty's position, it actually seems to reinstate a version of empiricism that neither Merleau-Ponty nor Dreyfus will endorse. Bodily motility cannot be understood along these oppositional lines of ideas and senses, just as habit and coping resist being understood as merely an empirical passive synthesis that is part of the actual. For them, our bodily comportment (including habits, coping, equilibrum and intentional arcs) is the place where the virtual and the actual do reciprocally determine one another.

25. Williams, *Deleuze's 'Difference and Repetition'*, 10.

26. In this respect, see the work of Dreyfus, Sean Kelly, Shaun Gallagher, and others, on the modification of neural networks in the brain and the way in which such a model resists the tacit philosophy of mind that is the basis for what is sometimes called "Good Old-Fashioned AI."

27. This is why Todd May's anarchist interpretation of poststructuralism has a certain plausibility. Todd May, *The Political Philosophy of Poststructuralist Anarchism* (Albany, NY: SUNY Press, 1997).

28. In *Discipline and Punish*, for example, Foucault famously describes the normalisation of social practices that takes place at the level of the body and has hence

shown us some of the reasons to be wary of habit, notably that "disciplinary coercion establishes in the body the constricting link between an increased aptitude and an increased domination." See Michel Foucault, *Discipline and Punish: The Birth of The Prison* (New York: Vintage, 1995), 138. In relation to this possibility, however, social and political analyses are called for (e.g,. are the practices in which our coping takes place worth preserving or do they need transforming?) rather than the a priori rejection of habituality and coping favoured by Deleuze and poststructuralism generally.

11. Touched by Time: Some Critical Reflections on Derrida's Engagement with Merleau-Ponty in *Le Toucher*

1.   An earlier version of the chapter was published in *Sophia* (2008), after being first presented at the International Association of Philosophy and Literature conference in Cyprus, 2007.

2.   Jacques Derrida, *Memoirs of the Blind: The Self-Portrait and Other Ruins*, trans. Pascale-Anne Brault and Michael Naas (Chicago: University of Chicago Press, 1993), 52.

3.   Ronald Bruzina describes Derrida's work "less as anti-presentialism, i.e. asserting that presence has neither validity nor significance, than as countering presentialism, i.e. marshalling ways of indicating presence is not an absolute secured by own manifestness, but rather a kind of 'constitutive' result, the 'constitution' in this case being the paradoxical systematically unmanifestable play of 'différancing.'" Ronald Bruzina, "The Future Past and Present—and Not Just Perfect—of Phenomenology", *Research in Phenomenology*, 30, 1 (2000): 51-2. Despite Derrida's suggestion in "Ousia and Gramme" that perhaps there is no concept of time that is not metaphysical, his transcendental arguments for the necessity of something that interrupts presentist time, and that opens on to the past and the future, amount to something rather close to a nonpresentist philosophy, or so I will argue in what follows.

4.   In *What Is Philosophy?*, Deleuze and Guattari make precisely the same comment about the concept of the flesh that features in Merleau-Ponty's later work, describing it as a "pious" thought that "plunges into the mystery of the incarnation" (178). Despite Merleau-Ponty's early Christianity and the problematic persistence of metaphors like "communion" and "original ecstasy" in his work (as Derrida shows), Deleuze and Guattari's reading of Merleau-Ponty's later ontology is nonetheless problematic. See Reynolds and Roffe, "Deleuze and Merleau-Ponty," 228-51.

5.   Derrida, *Speech and Phenomena*, 62-63.

6.   The findings of much contemporary cognitive science, for example, suggest that the history of the tradition, and particularly phenomenology, far from being wrong, may have provided an account of the human congruent with both phenomenology, the findings of developmental psychology and neonatal life. In particular, evidence suggests that there is an originary synaethesia of the senses (see Gallagher, *Body Shapes the Mind*, 160) despite the fact that Derrida bemoans the philosophical explication of this "confusion" or ambiguity in Merleau-Ponty's work (OT 193). Similarly, even Derrida's denigration of the privileged role given to the

hand is at least partially redeemed when it is recognized that this may be "hard-wired" into the human constitution—hand-mouth relations, for example, govern both foetal and neonatal life, and arguably this priority is never wholly abandoned. Whether this justifies these philosophical traditions depends upon what one takes the task of a philosophy to be, but it is perhaps not overly surprising (and it may even be desirable) that phenomenology serve to explicate experiential structures that are likely to have concomitant (but irreducible) explanations in the cognitive sciences, developmental sciences, etc.

7.   Thanks to Jon Roffe for helping to clarify the significance of the transcendental to this dispute and for proffering this likely reply.

8.   Robert Stern, *Transcendental Arguments and Skepticism* (Oxford: Clarendon Press, 2000), 10-11. He argues that truth directed transcendental arguments are ultimately problematic, relying on other question-begging assumptions, particularly either idealism or verificationism. But this is an epistemological reading of transcendental reflection, which does not necessarily jeopardize their role in regard to claims about meaning, nor as a regulative ideal.

9.   John Searle says that one of the main problems with phenomenological methods is that, using them, most of the important questions in philosophy and science cannot even be stated. See John Searle, "Neither Phenomenological Description Nor Rational Reconstruction: Reply to Dreyfus" (1999), 1, 10, http://socrates.berkeley.edu/~jsearle/articles.html (accessed November 12 2009). I do not think, however, that this is true of contemporary phenomenological practice.

10.   See, for example, Gallagher, *Body Shapes the Mind*, Dreyfus, "The Challenge of Merleau-Ponty's Phenomenology," and Kelly, "Grasping at Straws."

11.   These observations are partly indebted to Gayle Salamon. Her paper "The Sexual Schema: Transposition and Transcendence in Phenomenology of Perception," delivered at the International Association of Philosophy and Literature, Cyprus, 2007, reminded me of these aspects of Merleau-Ponty's work that have partly been forgotten due to Judith Butler's critique of his chapter on sexuality for its unproblematized assumption of gender neutrality.

12.   Derrida, *Adieu to Levinas*, 52.

13.   In both "Intersubjectivity: Notes on Merleau-Ponty" and "Sensibility", Levinas criticizes Merleau-Ponty's philosophy for being an imperialism of the same, and for being sustained by an unaccountable affection. While Levinas accepts Merleau-Ponty's descriptions of reversibility as they pertain to an individual touching themselves while touching another object—he describes it as a "remarkable analysis"—he is critical of the extending of this type of reversibility on to the alterity of another person. See Emmanuel Levinas, "Intersubjectivity: Notes on Merleau-Ponty," and "Sensibility," in *Ontology and Alterity in Merleau-Ponty*, eds. Galen Johnson and Michael Smith, trans. Michael Smith (Evanston, IL: Northwestern University Press, 1990), 55-66.

14.   Jacques Derrida, *Le Toucher: Jean-Luc Nancy* (Paris: Galilée, 2000), 218.

15.   In this respect, see David Morris, "The Enigma of Reversibility and the Genesis of Sense in Merleau-Ponty," *Continental Philosophy Review* 43, No. 2 (2010): 141-65.

16.   The last couple of pages of the pivotal chapter, "Tangent III," do, however, give the "Working Notes" for this book some attention.

17. Jacques Derrida, *Monolingualism of the Other or the Prosthesis of Origin*, trans. P. Mensah (Palo Alto, CA: Stanford University Press, 1996).

18. Thomas Fuchs, "Corporealized and Disembodied Minds: A Phenomenological View of the Body in Melancholy and Schizophrenia," *Philosophy, Psychiatry, Psychology* 12, 2 (2005): 98.

19. Andrew Meltzoff and Kevin Moore, "Imitation in Newborn Infants: Exploring the Range of Gestures Imitated and the Underlying-Mechanisms," *Developmental Psychology* 25 (1989): 954-62.

20. Again, however, Derrida would contest any too easy distinction between incorporation and introjection. See, for example, Jacques Derrida, "Fors: The Anglish Words of Nicolas Abraham and Maria Torok," trans. Barbara Johnson, in *The Wolfman's Magic Word: A Cryptonomy*, eds. Nicolas Abraham and Maria Torok (Minneapolis: University of Minnesota Press, 1986).

21. Recent analyses of brain functioning likewise suggest that bodily self-awareness and the perception of others share very closely related neurobiological functions. While the discovery of "mirror neurons" has perhaps had more made of it than the still incipient science justifies, it is interesting to note that when witnessing certain intentional acts like grasping neurons "fire" in both macaque monkeys and humans in the same ways as they fire when when the agent performs such actions themselves. There are, then, forms of embodied coupling with others, not only in utero but in neonatal perceptual and emotional life.

22. This point is well explored by James Hatley in his essay, "Recursive Incarnation and Chiasmic Flesh: Two Readings of Paul Celan's 'Chymisch'" in *Chiasms*, eds. Fred Evans and Len Lawlor (Albany, NY: SUNY Press, 2000), 237.

12. Heidegger and Derrida on Being-towards-death and Philosophy's Untimely Future

1. Gabriel Marcel, as cited in Geoffrey Scarre, *Death* (Montreal: McGill-Queen's University Press, 2007), 65.

2. Cristina Lafont, "Heidegger and the Synthetic A Priori," in *Transcendental Heidegger*, eds. Steven Crowell and Jeff Malpas (Cambridge, MA: MIT Press, 2008); 104-18.

3. This is Nicholas Rescher's understanding in *Aporetics* (Pittsburgh: University of Pittsburgh Press, 2009), 1.

4. Jacques Derrida, *On the Name*, ed. and trans. Tom Dutoit (Palo Alto, CA: Stanford University Press, 1995), 75.

5. Martin Heidegger, *The Fundamental Concepts of Metaphysics: World, Finitude, Solitude*. trans. William McNeill and Nicholas Walker (Bloomington: Indiana University Press, 1995), 185.

6. David Farrell Krell, *Heidegger: Basic Writings*, ed. David F. Krell (London: Blackwell, 1993), 22.

7. Levinas, *Time and the Other*, 71-72.

8. Alan Murray, "Philosophy and the Anteriority Complex," *Phenomenology and the Cognitive Sciences* 1, 1 (2002): 27-47.

13. Conclusion: Beyond Chronopathologies

1.   See the various essays contained in William McKenna and Jon Evans, eds. *Derrida and Phenomenology* (Dordrecht: Kluwer, 1995). Apart from the chapter by Len Lawlor, the other Husserlian phenomenologists maintain that Derrida is confused. See also the exchange between Kevin Mulligan and Tim Mooney. Mulligan's paper is titled, "How Not to Read: Derrida on Husserl," *Topoi* 10, 2 (1991): 199-208. Mooney's article is titled, "How to Read Once Again: Derrida on Husserl," *Philosophy Today* 47, 3 (2003): 305-21. Of course, the Mulligan and Mooney debate is perhaps overdetermined by the analytic-continental divide, with Mulligan on the side of the former and happy to consider Husserl an honorary analytic of sorts.

2.   Hans-Johann Glock, *What Is Analytic Philosophy?* (Cambridge: Cambridge University Press, 2008), 89-107.

3.   Thomas Akehurst, "The Nazi Tradition: The Analytic Critique of Continental Philosophy in Mid-century Britain," *History of European Ideas* 34 (2008): 557.

4.   John McCumber also makes a related point in *Reshaping Reason* (cf. 33).

5.   De Beauvoir, *Ethics of Ambiguity*, 91.

6.   Aristotle, *Nichomachean Ethics* (Harrmondsworth: Penguin, 2003), 1106b36-1107a3.

Bibliography

Akehurst, Thomas. "The Nazi Tradition: The Analytic Critique of Continental Philosophy in Mid-century Britain." *History of European Ideas* 34 (2008): 548-57.

Alliez, Éric. "Questionnaire on Deleuze." *Theory, Culture and Society*, trans. Philip Goodchild, and Nick Millett, 14, 2 (1997): 81-87.

———. *Capital Times: Tales From the Conquest of Time*. Trans. Georges Van Den Abbeele. Minneapolis: University of Minnesota Press, 1995.

Allison, David. "Derrida and Husserl." In *Understanding Derrida*. Eds. Jack Reynolds and Jon Roffe. London: Continuum, 2004, 113-21.

Ansell-Pearson, Keith. *Philosophy and the Adventure of the Virtual*. London: Routledge, 2002.

Aristotle. *Nichomachean Ethics*. Harmondsworth: Penguin, 2003.

Arnott, S. "The Problem of Solipsism and Deleuze's Ethics," *Contretemps* 2 (2001).

Badiou, Alain. *Being and Event*. Trans. Oliver Feltham. London: Continuum, 2005.

———. *Deleuze: The Clamour of Being*. Trans. Louise Burchill, Minneapolis: University of Minnesota Press, 1999.

———. "The Adventure of French Philosophy," *New Left Review* 35 (2005): 67-77.

———. "The Event in Deleuze," trans. Jon Roffe, *Parrhesia* 2 (2007): 37-44.

———. *Logiques des Mondes*. Paris: Editions de Seuil, 2006.

Beaney, Michael. "Analysis," *Stanford Encyclopedia of Philosophy*, http://stanford.library.usyd.edu.au/entries/analysis/

Beardsworth, Richard. "Jacques Derrida: The Power of Reason," *Theory and Event* 8, 1 (2005).

Beam, Gordon. "Differentiating Derrida and Deleuze," *Continental Philosophy Review* 33, 4 (2000): 441-65.

Benjamin, Andrew. *Style and Time: Essays on the Politics of Appearance*, Evanston, IL: Northwestern University Press, 2006.

Benjamin, Jessica. *The Bonds of Love*. New York: Pantheon Books 1988.

Bergson, Henri. *Duration and Simultaneity*. Ed. Robin Durie, Manchester: Clinamen Press, 1999.

———. *Matter and Memory*. Trans. N.M. Paul and W.S. Palmer, New York: Zone Books, 1994.

Blattner, William. *Heidegger's Temporal Idealism*. Cambridge: Cambridge University Press, 1999.

Boundas, Constantin. "Deleuze: Serialisation and Subject Formation." In *Deleuze and the Theatre of Philosophy*, Eds. Constantin Boundas and Dorothea Olkowski. New York: Routledge, 1994.

———. "Foreclosure of the Other: From Sartre to Deleuze," *Journal of the British*

261

Society of Phenomenology 24, 1 (1993): 32-41.

Bowden, Sean. "Deleuze et les Stoïciens: une logique of l'événement," *Bulletin de la Societe Américaine de Philosophie de Langue Française* 15, 1 (2005): 72-97.

Bradley, James. "Chapter 34: Transformations in Speculative Philosophy, 1914-45." *Cambridge History of Philosophy, 1870-1945*, Ed. Tom Baldwin, Cambridge: Cambridge University Press, 2003, 436-46.

Brandom, Robert. *Tales of the Mighty Dead*. Cambridge, MA: Harvard University Press, 2004.

Brown, Alan. *Modern Political Philosophy: Theories of the Just Society*. Harmondsworth: Penguin, 1986.

Bruzina, Ronald. "The Future Past and Present – and Not Just Perfect – of Phenomenology," *Research in Phenomenology* 30, 1 (2000): 40-53.

Calcagno, Antonio. *Badiou and Derrida: Politics, Events and their Time*. London: Continuum, 2007.

Campbell, Richard. "The Covert Metaphysics of the Clash Between Analytic and Continental Philosophy," *British Journal of the History of Philosophy* 9, 2 (2001): 341-59.

Caputo, John. *Deconstruction in a Nutshell*. New York: Fordham University Press, 1997.

———. *The Prayers and Tears of Jacques Derrida*. Bloomington: Indiana University Press, 1997.

Carr, David. *Phenomenology and the Problem of History*. Evanston, IL: Northwestern University Press, 1974.

———. *Time, Narrative and History*. Bloomington: Indiana University Press, 1991.

Chanter, Tina. *Time, Death and the Feminine: Levinas with Heidegger*. Palo Alto, CA: Stanford University Press, 2001.

Chase, James. "Analytic Philosophy and Dialogic Conservatism." In *Postanalytic and Metacontinental: Crossing Philosophical Divides*. Eds. Jack Reynolds, James Chase, James Williams, and Edwin Mares. London: Continuum, 2010, 85-104.

Chase, James, and Jack Reynolds. "The Fate of Transcendental Reasoning in Contemporary Philosophy." In *Postanalytic and Metacontinental: Crossing Philosophical Divides*. Eds. Jack Reynolds, James Chase, James Williams, and Edwin Mares. London: Continuum, 2010, 27-52.

———. *Analytic Versus Continental: Arguments on the Methods and Value of Philosophy*. Durham, UK: Acumen, 2010.

Christensen, Carleton B. "Getting Heidegger off the West Coast," *Inquiry* 41, 1 (1998): 65-87.

Cohen, Jonathan. *The Dialogue of Reason*. Oxford: Oxford University Press, 1986.

Connolly, William. *Politics of Ambiguity*. Wisconsin: University of Wisconsin Press, 1987.

Cooper, David. "Nietzsche and the Analytic Ambition," *Journal of Nietzsche Studies* 26, 1 (2003): 1-11.

Critchley, Simon. *Continental Philosophy: A Very Short Introduction*. Oxford: Oxford University Press, 2001.

———. *The Ethics of Deconstruction: Derrida and Lévinas*. Oxford, UK: Blackwell,

1992.
Crowell, Steven. "The Project of Ultimate Grounding and the Appeal to Inter-subjectivity in Recent Transcendental Philosophy," *International Journal of Philosophical Studies* 7, 1 (1999): 31-53.
Davies, Caroline. "As the Crow Flies: Follow the A34 and Turn Right". *The Age*, reprinted from the *Telegraph*.
de Beauvoir, Simone. *The Ethics of Ambiguity*. Trans. Bernard Frechtman. New York: Kensington Publishing, 1976.
De Bestegui, Miguel. *Truth and Genesis: Philosophy as Differential Ontology*. Bloomington: Indiana University Press, 2004.
De Landa, Manuel. "Deleuze, Diagrams, and the Open-ended Becoming of the World." In *Becomings; Explorations in Time, Memory, and Futures*. Ed. Elizabeth Grosz. Ithaca, NY: Cornell University Press, 1999, 29-41.
Deleuze, Gilles. "Nomad Thought". In *The New Nietzsche: Contemporary Styles of Interpretation*. Ed. David Allison. New York: Delta, 1977.
———. *Bergsonism*. Trans. Hugh Tomlinson and Barbara Habberjam. New York: Zone Books, 1988.
———. *Difference and Repetition*. Trans. Paul Patton. New York: Columbia University Press, 1994.
———. *Empiricism and Subjectivity*. Trans. Constantin Boundas. New York: Columbia University Press, 1991.
———. *Nietzsche and Philosophy*. Trans. Hugh Tomlinson. New York: Columbia University Press, 1983.
———. *Pure Immanence: Essays on a Life*. Ed. John Rajchman. Trans. Anne Boyman. New York: Zone Books, 2001.
———. *The Logic of Sense*. Trans. Mark Lester and Charles Stivale. London: Continuum, 2004.
———. "Coldness and Cruelty." In *Masochism*. Trans. Jean McNeil. New York: Zone Books, 1991.
———. "From Sacher-Masoch to Masochism," Trans. Christian Kerslake. *Angelaki: Journal of the Theoretical Humanities* 9, 1, (2004): 125-33.
Deleuze, Gilles, and Felix Guattari. *A Thousand Plateaus: Capitalism and Schizophrenia*. Trans. Brian Massumi. Minneapolis: University of Minnesota Press, 1987.
———. *Anti-Oedipus: Capitalism and Schizophrenia*. Trans. Richard Hurley, Mark Seem and Helen Lane. New York: Viking Press, 1977.
———. *What Is Philosophy?* Trans. Hugh Tomlinson and Graeme Burchill. London: Verso, 1994.
Derrida, Jacques. *Given Time: i. Counterfeit Money*. Trans. Peggy Kamuf. Chicago: University of Chicago Press, 1992.
———. "Fors: The Anglish Words of Nicolas Abraham and Maria Torok." In *The Wolfman's Magic Word: A Cryptonomy*, Eds. Nicolas Abraham and Maria Torok. Trans. Barbara Johnson. Minneapolis: University of Minnesota Press, 1986.
———. *Ear of the Other: Otobiography, Transference, Translation*. Trans. Peggy Kamuf. Ed. McDonald. London: University of Nebraska Press, 1985.
———. "Ousia and Gramme: A Note to a Footnote in *Being and Time*." In

Phenomenology in Perspective, Ed. Smith. Trans. Edward Casey. The Hague: Nijhoff, 1970.

———. *A Taste for the Secret*. Trans. G. Donis. London: Polity Press, 2001.

———. *Adieu to Emmanuel Levinas*. Trans. Pascale-Anne Brault and Michael Naas. Palo Alto, CA: Stanford University Press, 1999.

———. *Aporias*. Trans. Tom Dutoit. Palo Alto, CA: Stanford University Press, 1993.

———. *Deconstruction and the Possibility of Justice*. Eds. and Trans. Drucilla Cornell et al. New York: Routledge, 1992.

———. *Demeure: Fiction and Testimony*. Trans. Elizabeth Rottenberg. Palo Alto: Stanford University Press, 2000.

———. *Jacques Derrida: Acts of Literature*. Ed. Derek Attridge. New York: Routledge, 1992.

———. *Le Toucher: Jean-Luc Nancy*. Paris: Galilée, 2000.

———. *Limited Inc*. Ed. Gerald Graff. Trans. Samuel Weber. Evanston, IL: Northwestern University Press, 1998.

———. *Margins of Philosophy*. Trans. Alan Bass, Chicago: University of Chicago Press, 1982.

———. *Memoires: for Paul de Man*. Trans. Lindsay, Culler, Cadava, and P. Kamuf. New York: Columbia University Press 1989.

———. *Memoirs of the Blind: The Self-Portrait and Other Ruins*. Trans. Pascale-Anne Brault and Michael Naas. Chicago: University of Chicago Press, 1993.

———. *Monolingualism of the Other or the Prosthesis of Origin*. Trans. P. Mensah, Palo Alto, CA: Stanford University Press, 1996.

———. *Negotiations: Interventions and Interviews, 1971-2001*. Ed. and Trans. Elizabeth Rottenberg. Palo Alto, CA: Stanford University Press, 2002.

———. *Of Grammatology*. Trans. Gayatri Spivak. Baltimore: John Hopkins University Press, 1976.

———. *On Cosmopolitanism and Forgiveness*. Trans. Mark Dooley and Joe Hughes. London: Routledge, 2001.

———. *On the Name*. Ed. and Trans. Tom Dutoit. Palo Alto, CA: Stanford University Press, 1995.

———. *On Touching: Jean-Luc Nancy*. Trans. C. Irizarry. Palo Alto: Stanford University Press, 2005.

———. *Points . . . Interviews, 1974-1995*. Ed. Samuel Weber. Trans. Peggy Kamuf et al. Palo Alto, CA: Stanford University Press, 1995.

———. *Politics of Friendship*. Trans. George Collins. New York: Verso, 1997.

———. *Positions*. Trans. Alan Bass. London: Athlone Press, 1981.

———. *Psyche: Inventions of the Other, 1*. Ed. and Trans. Peggy Kamuf and Elizabeth Rottenberg. Palo Alto, CA: Stanford University Press, 2007.

———. *Rogues: Two Essays on Reason*. Trans. Pascale-Anne Brault and Michael Naas. Palo Alto, CA: Stanford University Press, 2004.

———. *Spectres of Marx: The State of the Debt, the Work of Mourning, and the New International*. Trans. Peggy Kamuf. New York: Routledge, 1994.

———. *The Gift of Death*. Trans. David Wills. Chicago: University of Chicago Press, 1995.

———. *The Work of Mourning*. Eds. and Trans. Pascale-Anne Brault and Michael

Naas. Chicago: University of Chicago Press, 2001.

———. *"Speech and Phenomena"and Other Essays on Husserl's Theory of Signs.* Trans. David Allison. Evanston, IL: Northwestern University Press, 1973.

Derrida, Jacques, and Anne Dufourmantelle. *Of Hospitality.* Trans. Rachel Bowlby. Palo Alto: Stanford University Press, 2000.

Dillon, Martin. *Merleau-Ponty's Ontology.* Bloomington: Indiana University Press, 1988.

Diprose, Rosalyn. *Corporeal Generosity: On Giving with Nietzsche, Merleau-Ponty and Levinas.* Albany, NY: SUNY Press, 2002.

Dolev, Yuval. *Time and Realism: Metaphysical and Antimetaphysical Perspectives.* Cambridge, MA: MIT Press, 2007.

Dowden, Bradley. "Time." *Internet Encyclopedia of Philosophy,* http://www.iep.utm. edu/t/time.htm.

Dreyfus, Hubert, and Stuart Dreyfus. "The Challenge of Merleau-Ponty's Phenomenology of Embodiment for Cognitive Science." In *Perspectives on Embodiment: The Intersections of Nature and Culture.* Eds. Homi Haber and Gail Weiss, London: Routledge, 1999, 103-120.

———. "The Ethical Implications of the Five-Stage Skill-Acquisition Model," *Bulletin of Science, Technology, and Society* 24 (2004): 251-674.

———. *Mind Over Machine.* New York: Free Press, 1982.

———. Hubert. "What Is Moral Maturity? A Phenomenological Account of the Development of Ethical Expertise," http://www.nuc.berkeley.edu/courses/ classes/E-124/Moral_Maturity_8_90.pdf

———. *Being-in-the-World.* Cambridge, MA: MIT Press, 1991.

———. *What Computers Still Can't Do: A Critique of Artificial Reason.* Cambridge, MA: MIT Press, 1997.

Duke, George, Elena Walsh, James Chase, and Jack Reynolds. "Postanalytic Philosophy: Overcoming the Divide?". In *Postanalytic and Metacontinental: Crossing Philosophical Divides.* Eds. Jack Reynolds, James Chase, James Williams and Edwin Mares. London: Continuum, 2010, 7-24.

Dummett, Michael. *The Origins of Analytical Philosophy.* London: Duckworth, 1993.

Ebertz, Roger. "Is Reflective Equilibrium a Coherentist Model?" in *Empirical Knowledge: Readings in Contemporary Epistemology.* Ed. Paul Moser. Totowa, NJ: Rowman and Allanfeld, 1986, 193-214.

Engel, Pascal. "Analytic Philosophy and Cognitive Norms," *Monist* 82, 2 (1999): 218-32.

Faulkner, Keith. *Deleuze and the Three Syntheses of Time.* New York: Peter Lang, 2006.

Foucault, Michel. "The Art of Telling the Truth." In *Critique and Power.* Trans. Alan Sheridan. Ed. M. Kelly, Cambridge, MA: MIT Press, 1994.

———. "Theatrum Philosophicum." In *Language, Counter-Memory, Practice.* Ed. Donald F. Bouchard. Ithaca, NY: Cornell University Press, 1977.

———. *Discipline and Punish: The Birth of the Prison.* Trans. Alan Sheridan. New York: Vintage, 1977.

———. *Foucault Live.* Ed. S. Lotringer, Trans. L. Hochroth and J. Johnston. New York: Semiotext(e), 1996.

Freud, Sigmund. *Beyond the Pleasure Principle.* Trans. James Strachey. London:

Hogarth Press, 1986.

———. *Three Essays on Sexuality*. Trans. James Strachey. New York: Basic Books, 1962.

Fricker, Miranda. "Epistemic Injustice and a Role for Virtue in the Politics of Knowing," *Metaphilosophy* 34, 1-2 (2003): 154-173.

Friedman, Michael. *A Parting of the Ways: Carnap, Cassirer and Heidegger*. New York: Open Court, 2000.

Fuchs, Thomas. "Corporealized and Disembodied Minds: A Phenomenological View of the Body in Melancholy and Schizophrenia." *Philosophy, Psychiatry, Psychology* 12, 2 (2005): 95-107.

Gallagher, Shaun, and Dan Zahavi. *The Phenomenological Mind*. London: Routledge, 2008.

Gallagher, Shaun. *The Inordinance of Time*. Evanston, IL: Northwestern University Press, 1998.

———. "The Place of Phronesis in Postmodern Hermeneutics", *Philosophy Today* 37 (1993): 298-305.

———. *How the Body Shapes the Mind*. Oxford: Oxford University Press, 2006.

Gasché, Rodolphe. *The Tain of the Mirror: Derrida and the Philosophy of Reflection*. Cambridge, MA: Harvard University Press, 1986.

Gaut, Berys. "Justifying Moral Pluralism." In *Ethical Intuitionism: Re-evaluations*. Ed. Philip Stratton-Lake. Oxford: Oxford University Press, 2003, 137-60.

Glendinning, Simon. "Argument All the Way Down: The Demanding Discipline of Non-Argumento-Centric Modes of Philosophy." In *Postanalytic and Metacontinental: Crossing Philosophical Divides*. Eds. Jack Reynolds, James Chase, James Williams, and Edwin Mares. London: Continuum 2010, 71-83.

———. "Reply to Reynolds," *International Journal of Philosophical Studies* 17, 2 (2009): 273-80.

———. *The Idea of Continental Philosophy*. Edinburgh: Edinburgh Univeristy Press, 2006.

Glock, Hans-Johann. *What Is Analytic Philosophy?* Cambridge: Cambridge University Press, 2008.

Goodman, Nelson. *Fact, Fiction and Forecast*. Cambridge, MA: Harvard University Press, 1983.

Grosz, Elizabeth. "Thinking the New: Of Futures Yet Unthought." In *Becomings; Explorations in Time, Memory, and Futures*. Ed. Elizabeth Grosz. Ithaca, NY: Cornell University Press, 1999, 15-28.

———. *The Nick of Time*. Sydney: Allen and Unwin, 2004.

———. *Time Travels*. Sydney: Allen and Unwin, 2005.

———. *Volatile Bodies: Towards a Corporeal Feminism*. Sydney: Allen and Unwin, 1994.

Haddad, Samir. "Derrida and Democracy at Risk," *Contretemps* 4 (2004): 29-38.

———. "Inheriting Democracy to Come", *Theory and Event* 8, 1 (2005).

Hallward, Peter. "Deleuze and the World of Others," *Philosophy Today* 41, 4 (1997): 530-44.

———. *Out of this World: Deleuze and the Philosophy of Creation*. London: Verso, 2006.

Hansen, Mark. "Becoming as Creative Involution? Contextualising Deleuze and Guattari's Biophilosophy," *Postmodern Culture* 11, 1 (2000).

Hardt, Michael. *Gilles Deleuze: An Apprenticeship in Philosophy*. London: UCL, 1993.

Hardt, Michael, and Antonio Negri. *Empire*. Cambridge, MA: Harvard University Press, 2000.

———. *Multitude*. New York: Penguin, 2005.

Hare, R. M. *Moral Thinking*. Oxford: Clarendon Press, 1981.

Hart, Kevin. *The Trespass of the Sign: Deconstruction, Theology and Philosophy*. Cambridge: University of Cambridge Press, 1989.

Hatley, James. "Recursive Incarnation and Chiasmic Flesh: Two Readings of Paul Celan's 'Chymisch.'" In *Chiasms*. Eds. Fred Evans and Leonard Lawlor. Albany, NY: SUNY Press, 2000, 237-49.

Hegel, G. W. F. *Phenomenology of Spirit*. Trans. A. V. Miller, J. N. Findlay. Oxford: Oxford University Press, 1979.

Heidegger Martin. *Basic Writings*. Ed. David F. Krell. London: Routledge, 1996.

———. *Being and Time*. Trans. John Macquarrie and Edward Robinson. London: SCM Press, 1962.

———. *Kant and the Problem of Metaphysics*. Trans. R. Taft. Bloomington: Indiana University Press, 1997.

———. *On Time and Being*. Trans. Joan Stambaugh. New York: Harper, 1972.

———. *The End of Philosophy*. Trans. Joan Stambaugh. New York: Harper and Row, 1973.

———. *The Fundamental Concepts of Metaphysics: World, Finitude, Solitude*. Trans. William McNeill and Nicholas Walker. Bloomington: Indiana University Press, 1995.

Holland, Eugene. *Deleuze and Guattari's Anti-Oedipus: An Introduction to Schizo-analysis*. London: Routledge, 1999.

Holmgren, Margaret. "Wide Reflective Equilibrium and Objective Moral Truth," *Metaphilosophy* 18 (1987): 116-24.

Honig, Bonnie. *Political Theory and the Displacement of Politics*. New York: Cornell University Press, 1993.

Honneth, Axel. *Fragmented World of the Social*. Ed. Charles W. Wright. Albany, NY: SUNY Press, 1995.

———. *The Struggle for Recognition: The Moral Grammar of Social Conflicts*. Trans. J. Anderson. Cambridge: University of Cambridge Press, 1996.

Honoré, Carl. *In Praise of Slow: How a Worldwide Movement Is Challenging the Cult of Speed*. London: Orion, 2004.

Hooker, Bradley, Elinor Mason, and Dale Miller. eds. *Morality, Rules and Consequences: A Critical Reader*. Lanham, MD: Rowman and Littlefield, 2000.

Hooker, Bradley. "Intuitions and Moral Theorising." In *Ethical Intuitionism: Re-evaluations*. Ed. Philip Stratton-Lake. Oxford: Oxford University Press, 2003, 161-83.

———. "Reflective Equilibrium and Rule Consequentialism." In *Morality, Rules, and Consequentialism: A Critical Reader*. Eds. Bradley Hooker, Elinor Mason, and Dale Miller. Lanham, MD: Rowman and Littlefield, 2000, 222-38.

———. *Ideal Code, Real World.* Oxford: Clarendon Press, 2000.

Horkheimer, Max. *Critical Theory.* London: Continuum, 1975.

Hoy, David. *The Time of Our Lives: A Critical History of Temporality.* Cambridge, MA: MIT Press, 2008.

Husserl, Edmund. "On the Phenomenology of the Consciousness of Internal Time." *Collected Works.* Ed. R. Bernet. Trans. J. Brough. Vol. 4. Dordrecht: Kluwer, 1991.

———. *Cartesian Meditations: An Introduction to Phenomenology.* Trans. D. Cairns. The Hague: Nijhoff, 1960.

———. *Ideas Pertaining to a Pure Phenomenology and to a Phenomenological Philosophy. First Book: General Introduction to a Pure Phenomenology.* Trans. F. Kersten. The Hague: Nijhoff, 1982.

———. *The Crisis of European Sciences and Transcendental Phenomenology.* Trans. David Carr. Evanston, IL: Northwestern University Press, 1970.

Hutto, Daniel, and Matthew Ratcliffe. Eds. *Folk Psychology Reassessed.* New York: Springer, 2007.

Jackson, Frank. "Thought Experiments and Possibilities," *Analysis* 69, 1 (2009): 100-109.

———. *From Metaphysics to Ethics: A Defence of Conceptual Analysis.* Oxford: Clarendon Press, 1998.

James, William. *The Will to Believe and Other Essays in Popular Philosophy.* New York: Dover Publications, 1956.

Johnson, Galen, and Michael Smith. eds. *Ontology and Alterity in Merleau-Ponty.* Evanston, IL: Northwestern University Press, 1990.

Jones, Graham, and Jon Roffe. eds. *Deleuze's Philosophical Heritage.* Edinburgh: Edinburgh University Press, 2009.

Kant, Immanuel. *Critique of Pure Reason.* Trans. W. Pluhar. Bloomington: Hackett, 1996

Keenan, Dennis. *Death and Responsibility: The Work of Levinas.* Albany, NY: SUNY Press, 1999.

Kelly, Sean. "Grasping at Straws: Motor Intentionality and the Cognitive Science of Skilled Behaviour." In *Essays in Honour of Hubert L. Dreyfus, Volume 2: Heidegger, Coping and Cognitive Science.* Eds. Mark Wrathall and Jeff Malpas. Cambridge, MA: MIT Press, 2000, 161-78.

———. "Merleau-Ponty and the Body." In *The Philosophy of the Body.* Ed. Michael Proudfoot. London: Blackwell, 2001, 62-76.

Kerslake, Christian. *Deleuze and the Unconscious.* London: Continuum, 2007.

Kierkegaard, Soren. *Concluding Unscientific Postscript to Philosophical Fragments.* Vol. 1. Trans. H. Hong, and E. Hong. Princeton: Princeton University Press, 1992.

Knight, Carl. "The Method of Reflective Equilibrium: Wide, Radical, Fallible, Plausible," *Philosophical Papers* 35, 2 (2006): 205-29.

Körner, Stephan. "The Impossibility of Transcendental Deductions," *The Monist* 51, 3 (1967): 317-31.

———. "Transcendental Tendencies in Recent Philosophy," *The Journal of Philosophy* 63 (1966): 551-61.

Kuusela, Oskari. "Transcendental Arguments and the Problem of Dogmatism," *Inter-*

national *Journal of Philosophical Studies* 16, 1 (2008): 57-75.

Kymlicka, Will. *Contemporary Political Philosophy*. Oxford: Clarendon Press, 1990.

La Caze, Marguerite. *The Analytic Imaginary*. Ithaca, NY: Cornell University Press, 2002.

Lafont, Cristina. "Heidegger and the Synthetic A Priori." In *Transcendental Heidegger*. Eds. Steven Crowell and Jeff Malpas. Cambridge, MA: MIT Press, 2008, 104-18.

Lambert, Gregg. "*Une Grande Politique*, Or the New Philosophy of Right?," *Critical Horizons* 4, 2 (2003): 177-97.

Lawlor, Leonard. "Derrida," *Stanford Encyclopedia of Philosophy*, http://plato.stanford.edu/entries/derrida/

———. "The End of Phenomenology: Expressionism in Deleuze and Merleau-Ponty," *Continental Philosophy Review* 31 (1998): 15-34.

———. "The Relation as the Fundamental Issue in Derrida." In *Derrida and Phenomenology*. Eds. William McKenna, and Jon Evans. Dordrecht: Kluwer, 1995.

———. *Husserl and Derrida: The Basic Problem of Phenomenology*. Bloomington: Indiana University Press, 2002.

Le Doeuff, Michele. *Hipparchia's Choice: An Essay Concerning Women, Philosophy etc.* Trans. T. Selous. Oxford: Blackwell, 1991.

Levinas, Emmanuel. "Intersubjectivity: Notes on Merleau-Ponty" and "Sensibility." In *Ontology and Alterity in Merleau-Ponty*. Eds. Galen Johnson and Michael Smith. Trans. Michael Smith. Evanston, IL: Northwestern University Press, 1990, 55-60, 60-66.

———. "Meaning and Sense." In *Collected Philosophical Papers*. Trans. Alphonso Lingis. The Hague: Martinus Nijhoff, 1977.

———. "Philosophy and the Idea of Infinity." In *Collected Philosophical Papers*. Trans. Alphonso Lingis. The Hague: Martinus Nijhoff, 1977.

———. *Time and the Other*. Trans. Richard Cohen. Pittsburgh: Duquesne University Press, 1987.

———. *Totality and Infinity*. Trans. Alphonso Lingis. Pittsburgh: Duquesne University Press, 1969.

Lewis, David. *Counterfactuals*. Cambridge, MA: Harvard University Press, 1973.

———. *On the Plurality of Worlds*. London: Blackwell, 2001.

Lloyd, Genevieve. "Fate and Fortune: Derrida on Facing the Future," *Philosophy Today* 43 (1999): 27–35.

Lyotard, Jean-François. *Differend: Phases in Dispute*. Trans. Georges Van Den Abbeele. Minneapolis: University of Minnesota Press, 1989.

———. *The Inhuman: Reflections on Time*. Trans. Geoffrey Bennington and Rachel Bowlby. Palo Alto, CA: Stanford University Press, 1991.

———. *The Postmodern Condition: A Report on Knowledge*. Trans. Geoffrey Bennington and Brian Massumi. Minneapolis: University of Minnesota Press, 1984.

Macadam, Ethan. "John Rawls at the End of Politics," *Angelaki: Journal of the Theoretical Humanities* 9, 3 (2004): 33-57.

Malpas, Jeff. "The Transcendental Circle," *Australasian Journal of Philosophy* 75, 1 (1997): 1-20.

————. *Heidegger's Topology*. Cambridge, MA: MIT Press, 2006.

Margaroni, Maria. "The Time of a Gift," *Philosophy Today* 48, 1 (2004): 49-62.

Markosian, N. "Time," *Stanford Encylopedia of Philosophy*, http://stanford.library. usyd.edu.au/entries/time/

Marx, Karl, and Friedrich Engels. *The Communist Manifesto*. Ed. John E. Toews, Boston: St Martin's Press, 1999.

Massumi, Brian. *A User's Guide to Capitalism and Schizophrenia: Deviations from Deleuze and Guattari*. Cambridge, MA: MIT Press, 1992.

Matthews, Richard. "Heidegger and Quine on the (Ir)relevance of Logic for Philosophy." In *A House Divided: Comparing Analytic and Continental Philosophy*. Ed. Carlos G. Prado. Atlantic Highlands, NJ: Humanity Books, 2003.

May, Todd. *The Political Philosophy of Poststructuralist Anarchism*. Albany, NY: SUNY Press, 1997.

McCumber, John. "The Temporal Turn in German Idealism," *Research in Phenomenology* 32 (2002): 44-59.

————. *Reshaping Reason: Towards a New Philosophy*. Bloomington: Indiana University Press, 2007.

————. *Time in the Ditch*. Evanston, IL: Northwestern University Press, 2001.

————. *Time and Philosophy*. Durham, Acumen, 2011.

McInerney, Peter. *Time and Experience*. Philadephia: Temple University, 1991.

McKenna, William, and Jon Evans, Eds. *Derrida and Phenomenology*. Dordrecht: Kluwer, 1995.

McTaggart, John. "The Unreality of Time." In *The Philosophy of Time*. Eds. Robin Le Poidevin and Murray MacBeath. Oxford: Oxford University Press, 1993, 23-34.

Meillassoux, Quentin. *After Finitude*. Trans. Ray Brassier. London: Continuum, 2008.

Mellor, Hugh. "The Unreality of Tense." In *Philosophy of Time*. Eds. Robin Le Poidevin and Murray MacBeath. Oxford: Oxford University Press, 1993.

Meltzoff, Andrew, and Kevin Moore. "Imitation in Newborn Infants: Exploring the range of Gestures Imitated and the Underlying-Mechanisms," *Developmental Psychology* 25 (1989): 954-62.

Mengue, Philippe. *Deleuze et la question de la democratie*. Paris: L'Harmattan, 2003.

Merleau-Ponty, Maurice. "The Child's Relations with Others." In *The Primacy of Perception and Other Essays*. Ed. J. Edie and Trans. W. Cobb, Evanston, IL: Northwestern University Press, 1964.

————. *Humanism and Terror: An Essay on the Communist Problem*. Trans. John O'Neill. Boston: Beacon Press, 1969.

————. *Phenomenology of Perception*. Trans. Colin Smith. London: Routledge, 2000.

————. *Signs*. Trans. Richard McCleary. Evanston, IL: Northwestern University Press, 1964.

————. *The Visible and the Invisible*. Trans. Alphonso Lingis. Evanston, IL: Northwestern University Press, 1964.

Midgley, Mary. "Duties to Islands." In *Environmental Ethics*. Ed. Robert Elliot, Oxford: Oxford University Press, 1995.

Mill, John Stuart. "On Liberty." In *Three Essays*. London: Oxford University Press, 1975.

Mohanty, J. N. "Method of Imaginative Variation in Phenomenology". In *Thought Experiments in Science and Philosophy*. Ed. T. Horowitz and G. Massey. Lanham, MD: Rowman and Littlefield, 1991.

Mooney, Timothy. "How to Read Once Again: Derrida on Husserl," *Philosophy Today* 47, 3 (2003): 305-21.

Morris, David. "The Enigma of Reversibility and the Genesis of Sense in Merleau-Ponty," *Continental Philosophy Review* 43, 2 (2010): 141-65.

———. *The Sense of Space*. Albany, NY. SUNY Press, 2004.

Mouffe, Chantal. *The Democratic Paradox*. London: Verso, 2000.

Mulgan, Timothy. *Understanding Utilitarianism*. Chesham, UK: Acumen, 2007.

Mulligan, Kevin. "How Not to Read: Derrida on Husserl." *Topoi* 10, 2 (1991): 199-208.

Murray, Alan. "Philosophy and the Anteriority Complex," *Phenomenology and the Cognitive Sciences* 1, 1 (2002): 27-47.

Negri, Antonio. *Time for the Revolution*. Trans. M. Mandarini. London: Continuum, 2003.

Newton-Smith, William. *The Structure of Time*. London: Routledge, 1980.

Nietzsche, Friedrich. *Beyond Good and Evil*. Trans. Walter Kaufmann. New York: Random House, 1966.

———. *On the Genealogy of Morals and Ecce Homo*. Trans. Walter Kaufman. New York: Vintage Books, 1969.

———. *The Gay Science*. Ed. Bernard Williams. Trans. Josefine Nauckhoff and Adrian Del Caro. Cambridge: Cambridge University Press, 2001.

Nozick, Robert. *Anarchy, State, and Utopia*. New York: Basic Books, 1974.

Olafson, Frederick. "Heidegger à la Wittgenstein or Coping with Professor Dreyfus," *Inquiry* 37, 1 (1994): 45-64.

Patton, Paul, and John Protevi. Eds. *Between Derrida and Deleuze*. London: Continuum, 2003.

Patton, Paul. *Deleuze and the Political*. London: Routledge, 1999.

Pippin, Robert. *Modernism as a Philosophical Problem*. Oxford: Blackwell, 1991.

Prado, Carlos G. *A House Divided: Comparing Analytic and Continental Philosophy*. Atlantic Highlands, NJ: Humanity Books, 2003.

Price, Huw. *Time's Arrow and Archimedes' Point*. Oxford: Oxford University Press, 1996.

Protevi, John. *Political Physics: Deleuze, Derrida and the Body Politic*. London: Athlone, 2001.

Putnam, Hilary. *Renewing Philosophy*. Cambridge, MA: Harvard University Press, 1992.

Quine, W. V. O. *Word and Object*. Cambridge, MA: MIT Press, 1960.

Rajchman, John. "Diagram and Diagnosis." In *Becomings; Explorations in Time, Memory, and Futures*. Ed. Elizabeth Grosz. Ithaca, NY: Cornell University Press, 1999, 42-54.

Rakowski, E. "The Future Reach of the Disembodied Will," *Politics, Philosophy and Economics* 4, No. 1 (2005): 91-130.

Rapaport, Herman. *Later Derrida: Reading the Recent Work*. London: Routledge, 2004.

Rawls, John. *A Theory of Justice*. Cambridge, MA: Harvard University Press, 1972.

————. *Justice as Fairness: A Restatement*. Cambridge: Belknap Press, 2001.

————. *Political Liberalism*. New York: Columbia Press, 1996.

Redding, Paul. *Analytic Philosophy and the Return of Hegelian Thought*. Cambridge: Cambridge University Press, 2007.

Reiff, Mark. "The Politics of Masochism," *Inquiry* 46, 1 (2003): 29-63.

Rescher, Nicholas. *Aporetics*. Pittsburgh: University of Pittsburgh Press, 2009.

————. *Philosophical Reasoning*. London: Blackwell, 2001.

Reynolds, Jack, and Jon Roffe. "Deleuze and Merleau-Ponty: Immanence, Univocity, and Phenomenology," *The Journal of the British Society of Phenomenology* 37, No. 3 (2006): 228-51.

Reynolds, Jack. "Chickening Out and the Idea of Continental Philosophy," *International Journal of Philosophical Studies* 17, 2 (2009): 255-72.

————. "Common Sense and Methodology: Some Metaphilosophical Reflections on Deleuze and Analytic Philosophy," *Philosophical Forum* 41, 3 (2010): 231-58.

————. "Deleuze and Dreyfus on *l'habitude*, Coping and Trauma in Skill Acquisition", *International Journal of Philosophical Studies* 14, 4 (2006): 563-83.

————. "Deleuze's Other-structure: Beyond the Master-slave Dialectic but at What Cost?", *Symposium: Canadian Journal of Continental Philosophy* 12, 1 (2008): 67-88.

————. "Derrida and Deleuze on Time and the Future," *Borderlands: New Spaces in the Humanities* 3, 1 (2004).

————. "Derrida, Friendship, and the Transcendental Priority of the 'Untimely,'" *Philosophy and Social Criticism* 36, 6 (2010): 663-676.

————. "Habituality and Undecidability: A Comparison of Merleau-Ponty and Derrida on the Decision," *International Journal of Philosophical Studies* 10, 4 (2002): 449–466.

————. "Negotiating the Non-negotiable: Rawls, Derrida, and the Intertwining of Political Calculation and Ultra-politics," *Theory and Event* 9, 3 (2006).

————. "Reply to Glendinning," *International Journal of Philosophical Studies* 17, 2 (2009): 281-7.

————. "Sadism and Masochism - A Symptomatology of Analytic and Continental Philosophy?" *Parrhesia* 1 (2006): 88-111.

————. "The Master-slave Dialectic and The 'Sadomasochistic Entity': Some Objections," *Angelaki: Journal of the Theoretical Humanities* 14, 3 (2009): 11-25.

————. "The Problem of Other Minds: Solutions and Dissolutions in Analytic and Continental Philosophy," *Philosophy Compass* 5, 4 (2010): 326-35.

————. "Time Out of Joint: Between Phenomenology and Poststructuralism," *Parrhesia* 9 (2010): 55-64.

————. "Touched by Time: Some Critical Reflections on Derrida's Reading of Merleau-Ponty in *Le Toucher*," *Sophia* 47, 3 (2008): 311-25.

————. "Transcendental Priority and Deleuzian Normativity: Reply to James Williams," *Deleuze Studies* 2 1 (2008): 101-8.

————. "Wounds and Scars: Deleuze on the Time (and the Ethics) of the Event," *Deleuze Studies* 1, 2 (2007): 144-66.

————. *Merleau-Ponty and Derrida: Intertwining Embodiment and Alterity*. Athens:

Ohio University Press, 2004.

———. *Understanding Existentialism*. Chesham, UK: Acumen, 2006.

Rifkin, Jeremy. *Time Wars: The Primary Conflict in Human History*. New York: Touchstone, 1987.

Roffe, Jon. "Deleuze." *Internet Encyclopedia of Philosophy*. Eds. James Fieser and Bradley Dowden.

———. "Hume." In *Deleuze's Philosophical Heritage*. Eds. Graham Jones and Jon Roffe. Edinburgh: Edinburgh University Press, 2009.

Rorty, Richard. *Philosophy and the Mirror of Nature*. Princeton, NY: Princeton University Press, 1981.

Rubenfeld, Jed. *Freedom and Time*. New Haven, CT: Yale University Press, 2001.

Russell, Bertrand. *Mysticism and Logic*. London: Allen and Unwin, 1917.

———. "Philosophy of Bergson," *Monist* 22 (1912): 321-47.

———. *History of Western Philosophy*. London: Routledge, 2004.

———. *Our Knowledge of the External World*. London: Routledge, 1993.

———. *The Problems of Philosophy*. London: Oxford University Press, 1912.

Sacks, Mark. "The Nature of Transcendental Arguments," *International Journal of Philosophical Studies* 13, 4 (2005): 439-60.

Sandel, Michael. *Democracy's Discontents*. Cambridge, MA: Harvard University Press, 1996.

———. *Liberalism and the Limits of Justice*. Cambridge: Cambridge University Press, 1982.

Sartre, Jean-Paul. *"No Exit" and three other plays*. trans. S. Gilbert and L. Abel, New York: Vintage, 1956.

———. *Being and Nothingness: An Essay in Phenomenological Ontology*. Trans. Hazel Barnes, London: Routledge, 1994.

———. *Notebooks for an Ethics*. Trans. David Pellauer. Chicago: University of Chicago Press, 1992.

———. *The Transcendence of the Ego*. Trans. F. Williams and R. Kirkpatrick. New York: Noonday Press, 1962.

Scarre, Geoffrey. *Death*. Montreal: McGill-Queen's University Press, 2007.

Searle, John. "Neither Phenomenological Description Nor Rational Reconstruction: Reply to Dreyfus." 1999.

Sellars, John. "An Ethics of the Event: Deleuze's Stoicism," *Angelaki: Journal of the Theoretical Humanities* 11, 3 (2006): 157-71.

Sherover, Charles. *Time, Freedom and the Common Good*. Albany, NY: SUNY Press, 1989.

Sidgwick, Henry. *The Methods of Ethics*. London: Hackett Publishing, 7th edition, 1981.

Singer, Peter. "Sidgwick and Reflective Equilibrium," *Monist* 58 (1972): 490-517.

———. "The Right to be Rich or Poor." In *Reading Nozick: Essays on Anarchy, State, and Utopia*. Ed. Jeffrey Paul. London: Basil Blackwell, 1981, 37-53.

———. *Practical Ethics*. Cambridge: Cambridge University Press, 1993.

Smith, Daniel W. "Critical and Clinical." In *Deleuze: Key Concepts*. Ed. Charles Stivale. Chesham, UK: Acumen, 2005.

———. "Deleuze and Derrida, Immanence and Transcendence." In *Between Derrida and Deleuze*. Eds. Paul Patton and John Protevi. London: Continuum,

2003, 46-66.

———. "Deleuze and the Question of Desire: Towards an Immanent Theory of Ethics," *Parrhesia* 2 (2007): 66-87.

Sorensen, Roy. *Thought Experiments*. Oxford: Oxford University Press, 1998.

Sosa, Ernest. "Consciousness of the Self and of the Present." In *Agent, Language and the Structure of the World*. Ed. J. Tomberlin. Bloomington: Hackett, 1983, 131-43.

Spinosa, Charles, Fernando Flores, and Hubert Dreyfus. *Disclosing New Worlds: Entrepreneurship, Democratic Action, and the Cultivation of Solidarity*. Cambridge, MA: MIT Press, 1999.

Stern, Robert. *Transcendental Arguments and Skepticism*. Oxford: Clarendon Press, 2000.

Stich, Stephen. *The Fragmentation of Reason*. Cambridge, MA: MIT Press, 1993.

Stiegler, Bernard. *Technics and Time, 1*. Trans. Richard Beardsworth and George Collins. Palo Alto, CA: Stanford University Press, 1998.

Stratton-Lake, Philip. Ed. *Ethical Intuitionism: Re-evaluations*. Oxford: Oxford University Press, 2003.

Strawson, Peter. *Individuals*. London: Routledge, 1990.

Stroud, Barry. "Transcendental Arguments," *Journal of Philosophy* 65 (1968): 241-56.

Stuhr, John. *Pragmatism, Postmodernism, and the Future of Philosophy*. London: Routledge, 1993.

Sutton, John. "Batting, Habit, and Memory: The Embodied Mind and the Nature of Skill," *Sport in Society* 10, 5 (2007): 763-86.

Swinburne, Richard. "Tensed Facts," *American Philosophical Quarterly* 27 (1990): 117-30.

Taylor, Charles. "The Validity of Transcendental Arguments." In *Philosophical Arguments*. Cambridge, MA: Harvard University Press, 1995, 20-33.

Taylor, D. "Phantasmatic Genealogy." In *Merleau-Ponty, Hermeneutics and Postmodernism*. Eds. Thomas Busch and Shaun Gallagher. Albany, NY: SUNY, 1992.

Thomson, Judith Jarvis. "A Defence of Abortion," *Philosophy and Public Affairs* 1, 1 (1971): 47-66.

Trebilcot, J. "Aprudentialism," *American Philosophical Quarterly* 11 (1974): 203-10.

Varzi, Achilles. Ed. "Time-Travel," *Monist* 88, 3 (2005).

Walzer, Michael. *Spheres of Justice: A Defence of Pluralism and Equality*. New York: Basic Books, 1984.

Weatherson, Brian. "What Good Are Counterexamples?," *Philosophical Studies* 115, 1 (2003): 1-31.

Weiner, Scott. "Inhabiting in the *Phenomenology of Perception*," *Philosophy Today* 34, 4 (1990): 342-353.

Weiss, Gail. "*Ecart*: The Space of Corporeal Difference." In *Chiasms: Merleau-Ponty's Notion of Flesh*. Eds. Fred Evans and Leonard Lawlor. Albany, NY: SUNY Press, 1995.

Wheeler, Samuel. *Deconstruction as Analytic Philosophy*. Palo Alto, CA: Stanford University Press, 2000.

White, Stephen. *Political Theory and Postmodernism*. Cambridge: University of

Cambridge Press, 1991.

Widder, Nathan. "Time Is Out of Joint - and So Are We: Deleuzean Immanence and the Fractured Self," *Philosophy Today* 50, 4 (2006): 411-31.

———. *Reflections on Time and Politics.* Pennsylvania, PA: Pennsylvania State University Press, 2008.

Williams, Bernard. *Truth and Truthfulness: An Essay in Genealogy.* Princeton, NJ: Princeton University Press, 2002.

———. *Utilitarianism: For and Against.* Cambridge: Cambridge University Press, 1976.

Williams, James. "Deleuze and J. M. W. Turner: Catastrophism in Philosophy." In *Deleuze and Philosophy.* Ed. Keith Ansell-Pearson. London: Routledge, 1997, 232-46.

———. "Deleuze and Whitehead: The Concept of Reciprocal Determination." In *Deleuze, Whitehead and the Transformation of Metaphysics.* Eds. A. Cloots and K. Robinson. Brussels: Konklijke Vlaamse Academie Van Belgie Voor Wetenschaapen En Kusten, 2005, 89-105.

———. "Why Deleuze Does Not Blow the Actual on Virtual Priority. A Rejoinder to Jack Reynolds", *Deleuze Studies* 2, 1 (2008): 97-100.

———. *A Critical Introduction and Guide to Gilles Deleuze's "Difference and Repetition."* Edinburgh: Edinburgh University Press, 2003.

———. *Gilles Deleuze's Philosophy of Time: A Critical Introduction and Guide.* Edinburgh: Edinburgh University Press, 2011.

———. *The Transversal Thought of Gilles Deleuze: Encounters and Influences.* Manchester: Clinamen Press, 2005.

Wiltshire, Bruce. *Fashionable Nihilism: A Critique of Analytic Philosophy.* Albany, NY: SUNY Press, 2002.

Wittgenstein, Ludwig. *Philosophical Investigations.* Trans. G. E. Anscombe. Oxford: Basil Blackwell, 1996.

Wood, David. *The Deconstruction of Time.* Atlantic Highlands, NJ: Humanities Press, 1989.

———. *The Step Back: Ethics and Politics after Deconstruction.* Albany, NY: SUNY Press, 2005.

Woodward, Ashley. "Jean-François Lyotard". In *Internet Encyclopedia of Philosophy,* Ed. James Fieser and Bradley Dowden.

Wrathall, Mark, and Jeff Malpas. *Essays in Honour of Hubert L. Dreyfus, Volume 2: Heidegger, Coping and Cognitive Science.* Cambridge, MA: MIT Press, 2000.

Young, Iris Marion. *Justice and the Politics of Difference.* Princeton, NJ: Princeton University Press, 1990.

Žižek, Slavoj. "Carl Schmitt in the Age of Post-Politics." In *The Challenge of Carl Schmitt.* Ed. Chantal Mouffe. London: Verso, 1999, 18-37.

Index